I0464507

QUANTUM CHEMISTRY

AUTHOR

Dr. K.P.Patel

(M.Sc., M.Ed., Ph.D., Chemistry)

(Principal)

Arts, Science and Commerce College
Lunawada Dist. Panchmahal,Gujarat (India)

PUBLISHED BY

The New Era International Publishing House
HQ. At & Po. Chaveli., Ta- Chansma,
Dist- Patan, North Gujarat, India, Asia.
www.iphouseindia.com

All rights reserved. Any person who does any unauthorized act in relation to this publication may be liable to criminal prosecution and civil claims for damages.

First Publication: 15th May, 2015

Copyright: Author

(c) **Dr. K.P.Patel**

ISBN:- 978-1-512-23771-9

Price: Rs.750/- INDIA

$ 15 OUTSIDE INDIA

PUBLISHED BY

The New Era International Publishing House
HQ. At & Po. Chaveli., Ta- Chansma,
Dist- Patan, North Gujarat, India, Asia.
www.iphouseindia.com

PREFACE

    This book has been written for the students of under- graduate and post graduate level of the various universities in India. A special feature of the book is that the text has been illustrated with a large number of line diagrams and the data presented in the form of numerous tables for reference and comparison. In the preparation of text standard works and review by renowned author have been freely consulted. At the end of the book will be found useful by those who wish to make a more details study of the topics discussed.

    However revealing of errors if any by the students and teachers which have missed attention and providing suggestions for further improvement of the book would be deemed a great favour not only to Author but also to the students community as a whole which would be gratefully acknowledged by the author.

    I feel proud and happy in remembering at this juncture all those whose inspiration , encouragement and co-operation have made my excellence book.

- Author

Dr. K.P.Patel

INDEX

Chapter-1
Starting of wave mechanics

Introduction :

The dynamic behaviour of macroscopic objects with which we come in to contact in daily life is governed by classical or Newtonian mechanics based on newton's three well-known laws of motion. It was only around the turn of the present century when phenomena like black body radiation, photoelectric effect, Compton effect and line spectrum of hydrogen atom were discovered that the inadequacy of classical mechanics was realized after a successful period of over three hundred years. Beginning with the revolutionary proposal of quantization of Max Planck in 1900 a new science governing the dynamical behaviour of subatomic constituents especially of electrons has evolved over a period of only about thirty years. This new science, known as quantum Mechanics, has been serving admirably in rationalizing chemical behaviout. Like any other theory or model, the quantum mechanics is also based on certain basic postulates. The only difference is that these postulates relate to atomic and molecular properties that are far from everyday experience and so are likely to appear strange and even unreasonable never the less these are to be taken on faith, the only justification being their ability to predict and correlate experimental facts. Again, quantum mechanics is an abstract mathematical model and calls for a great deal of tolerance and a good background of non-too-difficult mathematics to fully appreciate it. The practical results emerging from dry looking formalism are simply amazing.

There are two variants of quantum mechanics according to mathematical tools used. One is called "Matrix Mechanics" (Werner Heisenberg, 1925) based on use of matrix algebra while the other is called "Wave mechanics" (Ervin Schrodinger, 1926) which is based on classical wave theory and de Broglie's wave practicle duality. The latter being relatively simpler is more easily appreciated of the two. It is based on differential equations. However, what is referred to as quantum mechanics today is fusion of some aspects of both formalisms. It must be emphasized that in spite of the confirmed validity of the formalisms of quantum mechanics in measurements involving atomic structure, spectra, covalent bonds, free radicals, mechanisms mechnics still remains a model which we do not fully understand. However, the bewildering sequence of events beginning with Planck's quantum hypothesis in 1900 and culminating in quantum mechanical formalisms of Heisenberg and Schrodinger in 1926 represent a total recognization of the fundamental philosophy of science. In particular, they have forced an essential change in out attitude toward experimental observation. The rigid determinism of classical mechanics must give way to acceptance of the impossibility of experimentally determining or predicting the exact properties of an individual particle ar a given moment in time. Rather, we must deal with the statistical regularities which characterize a large number of measurements on a single particle. We must deal with probabilities rather than certainties. Quantum mechanics thus introduces a margin of uncertainity in our ability to measure and to know nature.

BLACK BODY RADIATION- PLANK'S QUANTUM HYPOTHESIS:

The prefect black body is one which obsorbs the heat radiation of whatever wavelength incident on it. It neither reflects nor transmits any of the incident radition and therefore appear black. But in nature no such body exists. But some body made up carbon obsorbs the most of the incident raditions. Radiation coming from the black body at a particular temperature shown in the fig.

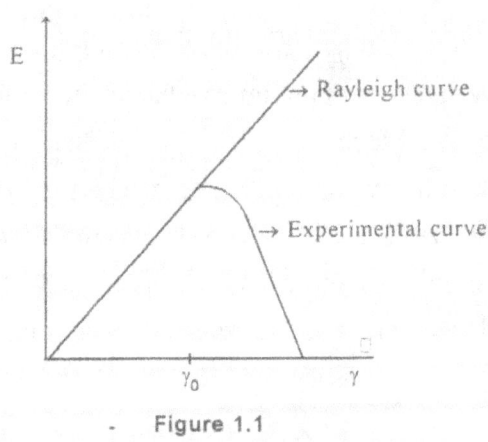

Figure 1.1

The Characteristic of the curve.

(i) energy is not uniformaly distributed in the radiation spectrum of the black body

(ii) intensity is maxium at a particular wave length, which is characteristic of the temperatures of the radiation body.

According to classical theory the average energy associated with each degrees of freedom is KT. Rayleigh-Jeans law deduced with help of classical theory says the energy per unit volume in the frequency range γ and $\gamma^+ d\gamma$ is given by According to this expression the energy emitted by a black body increases with the frequency and becomes infinite at large frequency. The experimentally obsorved radiation curve is complete disagreement with this conclusion. Since experimental curve shows that energy radiated at a particular temperature becomes zero both at lowar and higher frequency of shown in figure (1.1).

To clear this problem in 1900, Planck introduce a hypothesis known as Planck's hypothesis and according to this, "the black body radiation chamber is filled up not only with radiation but also the simple harmonic oscillator of molecular dimension known as Planck's oscillator which can vibrate the all possible frequency".

Plank assume that the energy of an oscillator of frequency various discreatly as that is multiples of small unit called quanta. This quanta of radiation is called photons. The energy of the photon is proportional to the frequency of the radiation that where h= 6.626 x 10-34 Jsec. The total energy of the oscillator which would be partly kinetic and partly potential must be = nh, where n= 0,1,2,3........ So the radiation would not be a continuous one, the discreat energies as O, h, 2h, ... Now the Planck radiation law deduced with the help of Planck quantum theory gives the energy desity interms of wave length belonging the range and as.toward experimental observation. The rigid determinism of classical mechanics must give way to acceptance of the impossibility of experimentally determining or predicting the exact properties of an individual particle ar a given moment in time. Rather, we must deal with the statistical regularities which characterize a large number of measurements on a single particle. We must deal with probabilities rather than certainties. Quantum mechanics thus introduces a margin of uncertainity in our ability to measure and to know nature.

HEISENBERG UNCERTAINTY PRINCIPLE :

This law of nature, first stated by Werner Heisenberg in 1927 emphasizes the fundamental difference between quantum mechanics and classical mechanics. One of the basic tenets of classical mechanics is that it is possible to know all dynamical variable of a system to any desired degree of accuracy. Such a strict determinism is untenable while dealing with quantum-mechanical systems (the microworld of atoms, molecules). In quantum mechanics we have to accept that there is a natural limit on the relative accuracy with which certain pairs of variables can be known. This was expressed as :

"precise measurements of the position and momentum of a particle like an electron

cannot be accoplished simultaneously."

Expressed mathematically : $x . px \geq h$

(h is the Planck constant)

If the position and momentum of a particle could be determined very precisely under quantum conditions then the particle-wave would have both a well-defined wavelength and a well-defined position. This seems contradictory. Imagine an experiment which simultaneoulsy determines the position and momentum of an electron. Say two precise measurements of position at different times are made from the elapsed time electron follows at once. However, in order to locate the electron with such precision it is necessary to "see" the electron by shining a radiation of extremely short wavelength i.e. photons of very large momentum on it. The interaction of the photon with the electron px)min would obviously change the momentum of the electron. Thus the very act of measuring its precise position renders its momentum imprecise. h represents the natural limit in the simultaneous precision px)min of position and momentum, i.e. $\left(x . \approx h \right)$

Note that :

(1)   the uncertainty principle is of no consequence in the macroworld due to very small magnitude of h.

(2)   Since x is related to wave description and px to particle description, the uncertainty principle emphasizes the mutual exclusiveness of these two aspects

which are to be regarded complementary.

(3)   Not all measurements are subject to this fundamental restriction. The uncertainty principle applies to canonically conjugate pairs of variables, like energy (E) and time (t)

or angular momentum (P) and angle (θ_θ). Thus

$$E . t \geq h$$

$$\theta . p_\theta \geq h$$

(4)   The uncertainty principle as stated here is a specific case of the generalized form which can be derived from quantum theory for any pair of variables, A and B. The generalized form of Heisenberg uncertainty principle is

$$A . B \geq \left| \frac{\left| A , B \right|}{2i} \right|$$

If A is $\$_x$ and B is P_x,

$$x , P_x = i\mathsf{H} \qquad \text{(prove yourself)}$$

$$\therefore \quad \$_{x , P_x} = i\mathsf{H}$$

$$\therefore \quad x . P_x \geq \left| \frac{i\mathsf{H}}{2i} \right|$$

$$\therefore \quad \left\langle x . \right\rangle P_x \geq \frac{\mathsf{H}}{2}$$

INTERPRETATION OF ψ :

If must be noted that in using $\psi|^2$ to represent "electron (particle) density" in quantum mechanics. We have drawn analogy with classical electrodynamic theory where

the energy density is represented in terms of field vectors H and E as

$$E = \frac{1}{8\pi} \int\int\int_{}^{R} |z|^2 \, dxdydz$$

Where Z is a new field vector defined as

$$Z = E + iH$$

Moreoever ψ shows a typical behaviour of a vector. Just as a vector can be resolved in to components.

$$A = i\,A_x + j\,A_y + k\,A_z$$

Similarly, any arbitrary state ψ of a system can be expressed as a linear combination

of eigen states, φ_i

$$\psi = a_1\,\varphi_1 + a_2\,\varphi_2 + \ldots\ldots\ldots\ldots \sum_n a_n\,\varphi_n$$

Following Dirac ψ may be treated as a vectore (called the state vector) and represented as 'Ket ψ', $\psi \rangle$.

The associated dual vectore (equivalent of the complex conjugate ψ*) is called the 'bra ψ' and is represented as $\langle \psi$. In this Dirac notation, the combination of bra and ket--- *bra(c)ket - represents an intergral e.g.,*

$$\langle \psi | \psi \rangle = \int \psi^* \, \psi \, d\tau$$

$$\langle \psi | H | \psi \rangle = \int \psi^* \, H \, \psi \, d\tau$$

In quantum mechanics, the length of a ket vector has no physical meaning (i.e., multuplication by a scalar does not alter the state; if ψ is a solution, then aψ is also a solution); only the direction matters.

THE SCHRODINGER EQUATION   ($H\Psi = E\Psi$)

Stable wave function has node & Antinode. Hight of wave is increased from node to antinode. While it is decreased from Antinode to node. So changing of energy with the changing of particle. We call it wave function Ψ. Ψ is depend on time. But for the system of atom Ψ is free from atom so.

$\Psi = A \sin wt$

$\Psi = A \sin \frac{2\pi}{\tau} .t$ (τ જયાં આવર્તકાળ એક ચક્ર પૂર્ણ કરવા લાગતો સમય $w = \frac{2\pi}{\tau}$)

$\Psi = A \sin 2\pi . \frac{t}{\tau}$

$\Psi = A \sin 2\pi . \frac{x}{\lambda}$ ....(A) (આવર્તકાળ τ = તરંગલંબાઈ $\therefore t = ?$, $\therefore \frac{t\lambda}{\tau} = x$ ધારો, $\therefore \frac{t}{\tau} = \frac{x}{\lambda}$)

Differentiation of Eq (A)

$\frac{d\Psi}{dx} = \frac{2\pi}{\lambda} A \cos \frac{2\pi x}{\lambda}$

and $\frac{d^2\Psi}{dx^2} = -\frac{4\pi^2}{\lambda^2} . A \sin \frac{2\pi x}{\lambda}$

$$\frac{d^2\Psi}{dx^2} = -\frac{4\pi^2}{\lambda^2}.\Psi \ \ldots\ldots\ldots\ldots \ (B)$$

Total energy of particle is given by T + V

$\therefore E = T + V$

wher, $T = \frac{1}{2} mv^2$

$$T = \frac{m^2 v^2}{2m} = \frac{P^2}{2m} \implies P^2 = 2Tm \ (\ P = mv \ \text{વેગમાન} \ ) \ \ldots\ldots\ldots (C)$$

By the eq. or De Broqlie

$$\lambda = \frac{h}{p}$$

$$\therefore \frac{h}{\lambda}$$

$$\therefore P^2 = \frac{h^2}{\lambda^2} \ \ldots\ldots\ldots\ldots \ (D)$$

relationship between (C) + (D)

$$2Tm = \frac{h^2}{\lambda^2}$$

$$\therefore \lambda^2 = \frac{h^2}{2Tm} \ \ldots\ldots\ldots\ldots (E)$$

put the value of (E) in Eq. (B)

$$\frac{d^2\Psi}{dx^2} = \frac{-4\pi^2 \times 2Tm}{h^2}.\Psi$$

$$\frac{d^2\Psi}{dx^2} = \frac{-8\pi^2 m}{h^2} \times T \times \Psi$$

Now, $E = T + V \implies T = E - V$

$$\frac{d^2\Psi}{dx^2} = \frac{-8\pi^2 m}{h^2} (E-V).\Psi$$

$$\frac{d^2\Psi}{dx^2} + \frac{8\pi^2 m}{h^2} (E-V) \ \Psi = 0 \ \ldots\ldots\ldots\ldots \ (F)$$

(F) is the Schrödinger eq. for a single particle in one dimensions for a single particle in three dimensions is.

$$\frac{d^2\Psi}{dx^2} + \frac{d^2\Psi}{dy^2} + \frac{d^2\Psi}{dz^2} + \frac{8\pi^2 m}{h^2} (E-V) \ \Psi = 0$$

$$(\frac{d^2}{dx^2} + \frac{d^2}{dy^2} + \frac{d^2}{dz^2}) \ \Psi + \frac{8\pi^2 m}{h^2} (E-V) \ \Psi = 0$$

$$\nabla^2 + \frac{8\pi^2 m}{h^2} (E-V) \ \Psi = 0 \ \text{where } \nabla = \text{Laplasian operation multiply by } -\frac{h^2}{8\pi^2 m}$$

$$[-\frac{h^2}{8\pi^2 m} \nabla^2 + V] \ \Psi = E \ \Psi \ \ldots\ldots\ldots\ldots\ldots (G)$$

$$\boxed{H\Psi = E\Psi} \qquad \ldots\ldots (H) \ \text{where, } H = \text{Hemiltonian operator.}$$

Eq. (H) is the short form of Schrodinger equation.

TIME DEPENDENT SCHRODINGER WAVE EQUATION

કવોન્ટમ મિકેનિસકના સમયના પદનો સમાવેશ અતિ મહત્વનો છે. કોઈ પણ ક્ષણે ભૌતિક પ્રણાલી માટે તરંગ વિધેય Ψ લેવામાં આવે તો તે બિંદુએ પ્રણાલીના ફકત ગુણધર્મોનું જ વર્ણન કરતુ નથી. પરંતુ પ્રતિક્ષણે તે પ્રણાલીનું વર્તન પણ જણાવે છે. આ હકીકતને ગાણિતીય રીતે $\frac{d\Psi}{dt}$ વડે રજૂ કરવામાં આવે છે. જે t સમય માટે છે.

કવોન્ટમ યંત્રશાસ્ત્રની ચોથી અભિધારણા પ્રમાણે (4th According to postulate of quantum mechanism) તરંગ વિધેયની સમય સાથે વૃદ્ધિ નીચેના શ્રોડિંજર સમી. વડે રજૂ કરવામાં આવે છે.

$$\Psi_{(r,t)} = \Psi_{(r)} . \exp(-\frac{2\pi i}{h} E(t)) \ \ldots\ldots\ldots\ldots (1)$$

$$\text{where } \{ \ \Psi_{(t)} . \exp(-\frac{2\pi i}{h} E(t))$$

Eq....(1) નું વિકલન કરતાં (detrantial of Eq .....(1) is)

$$E \ \Psi = \hat{H} \ \Psi = \frac{ih}{2\pi} \frac{d\psi}{dt} \ \ldots\ldots\ldots\ldots (2) \ \{ \ \hat{H} = \text{Hemeltonian operator)}$$

Now, $\bar{H} = -\frac{h^2}{8\pi^2 m} \nabla^2 + V(r)$ ....(3)(a)

pnt the valu of $\bar{H}$ in Eq ....(2)

$[-\frac{h^2}{8\pi^2 m} \nabla^2 + V(r)]\Psi = \frac{ih}{2\pi} \frac{d\Psi}{dt}$ ....(3)

હવે $\nabla^2 = \frac{d^2}{dr^2}$ લખતાં

$-\frac{h^2}{8\pi^2 m} \frac{d^2\Psi}{dr^2} + \frac{ih}{2\pi} \frac{d\Psi(r,t)}{dt}$ ....(4)

તરંગ વિધેય Ψ ને r અને t ઉપર આધારિત બે તરંગ વિધેયો વડે રજૂ કરી શકાય છે. આ તરંગવિધેયો અનુક્રમે ψ_r અને ψ_t છે. તેથી Ψ ને બે સ્વતંત્ર અસ્તિત્વ ધરાવતાં તરંગ વિધેયના સાદા ગુણાકારથી દર્શાવી શકાય .

$\Psi_{(r,t)} = \Psi_{(r)} \Psi_{(t)}$ ....(5)

r ને અનુલક્ષીને સમી. (5) નું વિઘટન કરતાં

$\frac{d\Psi(r,t)}{dr} = \Psi'_{(t)} \frac{d\Psi(r)}{dr}$ ....(6)

r ને અનુલક્ષીને સમી. (6) નું ફરીથી વિકલન કરતાં

$\frac{d^2\Psi(r,t)}{dr^2} = \Psi'_{(t)} \frac{d^2\Psi(r)}{dr^2}$ ....(7)

તેવી જ રીતે સમી. (5) નુ t ને અનુલક્ષીને વિકલન કરતાં,

$\frac{d\Psi(r,t)}{dt} = \Psi_{(r)} \frac{d\Psi'(t)}{dt}$ ....(8)

સમી. (5), (7) & (8) ની કિંમતો સમી (4) માં મુકતા,

$-\frac{h^2}{8\pi^2 m} \cdot \Psi'_{(t)} \frac{d^2\Psi(r)}{dr^2} + V_r. \Psi_r \Psi'_{(t)} = \frac{ih}{2\pi} \Psi_r \frac{d\Psi'(t)}{dt}$

સમી. (9) ને $\Psi r\ \Psi'(t)$ વડે ભાગતાં,

$$\frac{-\hbar^2}{8\pi^2 m}\ \frac{1}{\Psi(r)}\ \frac{d^2\Psi(r)}{dr^2} + V_r = \frac{ih}{2\pi}\ \frac{1}{\Psi'(t)}\ \frac{d\Psi'(t)}{dt} \quad \ldots\ldots\ldots\ldots(10)$$

સમી. (10) માં ડા.બા.નું પદ માત્ર r પર આધારિત છે. જમણી બાજુનુ પદ t ઉપર આધારિત છે. પરિણામે બન્ને પદનું અલગીકરણ થાય છે. સમી.ની બન્ને બાજુઓને અચળાંક "E" સાથે સરખાવતાં બે અલગ સમી. મળે છે જે નીચે મુજબ છે.

(1) $-\dfrac{\hbar^2}{8\pi^2 m}\ \dfrac{d^2\Psi(r)}{dr^2} + V_r\Psi r = E\Psi(r)$ ....(11)

(2) $\dfrac{ih}{2\pi}\ \dfrac{d\Psi'(t)}{dt} = E\Psi'(t)$ ....(12)

Eq ---(11) એ સમયથી સ્વતંત્ર સમી. છે. જયારે સમી. (12) સમય આધારિત છે.

EIGEN VALUES AND EIGEN FUNCTIONS

If we implies operator $\hat{A}$ on any any wave function , Resulting this wave function back with multifunction of any constant values, we say this function is Eigen function and equation is Eigen equation which write as under.

$$\boxed{\hat{A}\ \Psi = a\ \Psi}$$ --------- (1)

(Where, Ψ = Eigen Function
a = Eigen Value
and Eq – (1) is Eigen Equation.)

Schrodinger equation $H\Psi = E\Psi$ is also eigen equation. Where $\bar{H}$ operator, Ψ eigen function n & E eigen value. If the value of Ψ gives the acceptable solution of schrodinger equation which known as a eigen function and resulting the total value of E by the satisfy solution of schrodinger equation which known as a eigen value for the system in quantum mechanics.

NORMALIZATION AND ORTHOQONALITY

Information of any system of moving particle $\bar{e}$ gets by the wave function Ψ value of $\Psi^2 dv$, gives the probability to finding the $\bar{e}$ in small value dv of any place of system. Now if $\bar{e}$ (particle) is in any system the probability to finding the $\bar{e}$ in system is sure. (ઈ મળવાની સંભાવના ચોક્કસ હોય છે.)

H-Atom has on $\bar{e}$. So, $\bar{e}$ is in the area of H- Atom therefore probability to finding the $\bar{e}$ in this area will be one. Value of probability of $\bar{e}$ in small value dv in various place of this area is known by $\Psi^2 .dv$. So, integral of $\Psi^2 .dv$. fof al area gives the value of probability to finding the $\bar{e}$ in the area of this system.

The integral of this wave function over the entire space in the box must be equal to unity because there is only one electron and at all times it is some where in the box . Therefore

$$\int_0^a I \ \Psi^2 n \ I \ dx = 1$$

If any wave function follow this condition known as normalized wave function. If Ψ is not normalized at all times multiply by N'

$$\therefore \int (N \cdot \Psi)^2 \ dv = 1 \qquad\qquad \text{where, N = Normalized constant}$$

$$\therefore N^2 \int (\Psi^2 \ dv = 1$$

$$\therefore \int (\Psi^2 \ dv = \frac{1}{N^2}$$

Consider the normalized wave function Ψn and $\Psi n'$ corresponding to two different states of an electron in a box, it is found that for $n \neq n'$

$$\int_0^a \Psi n \ \Psi n' \ dx = 0$$

The wave functions for different states of this system are thus orthogonal. This property of orthogonality between any two different states ensures that the various states are truly independent of one another.

Chapter-2
The Basic Postulates of Quantum Mechanics:

First Postulates of Quantum Mechanics

(a) The state of a dynamical system is described as fully as possible by a function ψ which is a function of all the coordinates qi of the particles of the system and of time. i.e. ψ = ψ (qi, t). This function is called the state function or the wave function.The concept of ψ is the single most fundamental and extremely important concept in quantum mechanics. It contains every possible information on the state described by it. The use of the phrase "as fully as possible" emphasises the natural limit to the knowability of a quantum mechanical system as embodied in the Heisenberg's uncertainty principle.

It is to be noted that ψ is generally complex. i.e., contains i = −1 and therefore, unlike thermodynamic state functions or more precisely their differences, cannot be measured by direct experiment. (This need not disturb us since properties depend on ψ* ψ which is always real. ψ* is the complex conjugate of ψ and is obtained by changing the sign of i where it occurs in ψ). ψ is a function of all the coordinates of all the particles present in the system and of time.

$$\psi = \psi(q_i, t) \qquad q_i = \text{generalized coordinates (s) of } i^{th} \text{ particles.}$$

Thus for a single particle, one dimensional (x) system, ψ = ψ(x, t) and for a single-particles three-dimensional system ψ = ψ(x,y,z,t). For the general system of n-particles,ψ = ψ(x, y, z, ... x_n, y_n, z_n, t). Consider ψ(x, t). The value of space coordinate x defines the location of the particle (or equivalently a particular configuration of the system) at time t. The permissible range of valies of x, sat $-\infty \leq x \leq +\infty$, defines all possible configurations of or the configuration space available to the system. In the general case, specifying, at a given instant t, all the space coordinates of all the n particles. viz. x_1 y_1 z_1 x_2 y_2 z_2 .... xn yn zn (a total of 3n coordinates) defines the cofiguartion (distribution of constituent particles) at the particular time t and specifying the total range of values of each coordinates, say

$$-\infty \leq x \leq +\infty \qquad\qquad i = 1, 2, ... n$$
$$-\infty \leq y \leq +\infty$$

$$-\infty \leq z \leq +\infty$$

defines the configuration space available to the system.

It has been said earlier that we have to be content with probabilistic descriptions while dealing with quantum mechanical systems and this forms the substance of the remainder of Postulate-1.

(b)   The nature of the function ψ is such that $\psi^* \psi\, d\tau$ is the probability that the system has a configuration described by ψ at time t. The state function ψ must be "well-behaved" i.e., finite, single-valued, continuous and quadratically integrable (in other words, a Class Q function). note that, by postulate, ψ is connected with probability, though not directly but through ψ*ψ. Since whereas ψ is generally complex (probability must be real) ψ*ψ is always real.

According to P.1(b), if ψ = ψ(x, t) then ψ*ψdx represents the probability of finding the particle at x (i.e. its coordinate between x and x + dx) at time t. Similarly, for a many - particle system with

$$\psi = \psi(q_i, t) = \psi(x_1\, y_1\, z_1\, x_2\, y_2\, z_2 \ldots\ldots\ldots\ldots\ldots\ldots x_N\, y_N\, z_N, t) \qquad \text{----- (4)}$$

$\psi^* \psi\, dx_1\, dy_1\, dz_1 \ldots\ldots\ldots\ldots dz_N = \psi^* \psi\, d\tau$ represents the probability that, at time t, particle 1 is found in a small volume element dV_1 at $(x_1\ y_1\ z_1)$ (i.e. with its x-coordinate between x_1 and $x_1 + dx_1$, y-coordinate between y_1 and $y_1 + dy_1$ and z-coordinate between z_1 and $z_1 + dz_1$, small volume element dV_2 at $(x_2\ y_2\ z_2)$, particle-3 in a small volume element dV_3 at $(x_3\ y_3\ z_3)$ etc.

It is to be noted that the quantum mechanical probability is to be interpreted in a statistical sense. Thus, operationally, $\psi^* \psi\, d\tau$ represents the relative number of times in a large number of measurements that the system is found in a particular configuration descrived by ψ at time t. This is the Born interpretation of ψ and $\psi^* \psi$ is called the probability density or distribution function. The latter part of P.1(b) concerning restriction on well-behaved ψ automatically follow since ψ must be in accord with physical reality. Since the probability cannot be allowed to be infinite (infinite probability would mean certainty and the uncertainty principle will be violated) and has to have just one value at a given point, the requirements that ψ be finite and single-valued are obvious. Continuity requirement follows similarly. By square integrable it is meant that integrated probability must equal a finite number i.e.

$$\int \int \dots\dots \int \psi^* \psi \, d\tau$$

= Q (overall

configuration space)

obviously since all the particles must lie somewhere in the configuration space the integrated probability density must be unity, i.e.,

$$\int \psi^* \psi \, d\tau = 1$$

all space

The function ψ is then said to be normalized (to unity). We shall generally work with normalized wave function.

Note : $\psi^* \psi = \psi \psi^* = |\psi|^2 = P$ is also interpreted as "electron density" in atomic and molecular problems. If $\psi = \psi(x)$ $\psi(x)^2 = P(x)$ represents the number of electrons (or a fraction of an electron in a single electron case) present between x and x + dx.

SECOND POSTULATES OF QUANTUM MECHANICS

If A is and operator associated with an observable A and there is a set of identical systems known to be in the state ψ_n that is an eigenfunction of A, i.e.,

$$A\psi_n = a_n \psi_n \qquad (a_n \text{ is a number})$$

then, if an experimentalist makes a series of measurements of the property A, it will get the same result a_n. It is only under this condition that an experiment will give a precise result." OR

"The only possible values that can be observed for a dynamical variable in an assembly of systems are the eigenvalues a_i in the equation.

$$A\psi_i = a_i \psi_i$$

Where A is the operator for the variable and ψ_i is any well-behaved eigenfunction of A.

(vi)   Expectation value of an operator : (mean velue theorem)

if ψ is not an eigenfunction of A,

$$A\psi \neq a\psi = \psi'$$

The operation by the operator (equivalent to an attempt to experimentally measure the property A) changes the state and so repeated measurements will not give the same result. Thus the property A in such a case is note well-defined (or sharp; it cannot be determined with complete certainty. In a series of measurements on identical systems, we shall get a range of values. where a precise value is impossible to obtain, we shall have to be content with only the mean or the average value (the expectation value of the operator) obtained as stated in P.2(b).

(b)   **MEAN VALUE THEOREM**

When a great many measurements of any dynamical variable A are made on an assembly of identical systems all described by a function ψ that is not an eigenfunction

of the operator A associated with the vatiable A, a distribution of results will be obtained. Although we cannot predict the result of any particular measurement, the average or mean value of A,

$$\langle A \rangle = \overline{A} \text{ will be given by}$$

$$\langle A \rangle = \overline{A} = \int \psi * A \psi \, d\tau \, / \, \psi * \psi \, d\tau \qquad \text{(general)}$$

$$= \int \psi * A \psi \, d\tau \qquad \text{(ψ normalized)}$$

Is is easy to show that the expectation value is same as the eigenvalue if ψ is an

---

16

eigenfunction of A.

Let $A\psi = a\psi$ and $\int \psi^*\psi d\tau = 1$ (ψ normalized)

$$\therefore \langle A \rangle = \overline{A} = \int \psi^* A \, \psi \, d\tau = \int \psi^* a \, \psi \, d\tau = a \int \psi^* \psi \, d\tau = a$$

OPERATORS :

In order to extract the information contains in ψ it is necessary to "operate" on it. We require operators. The concept of an operator is introduced in P.2.

"For every dynamical variable there must be assigned a linear, Hermitian operator. The physical properties of the variable can be deduced from the mathematial properties of the associated operator."

We have been using operators all along (e.g. d/dx, $\sqrt{}$, $()^2$, etc.) The only thing is that we have not thought about them apart from the functions on which they operate.

An operator may be defined as "a symbol which represents a well-defined set of mathematical operations to be carried out on a function (operand) that follows it."

Alternatively, an operator may be defined as "a rule of changing one function in to another". Symbolically.

$$O \psi = \psi'$$

(O stands for a general operator. Circumflex is used over the letter to indicate that it is an operator) Note that operators without their operands are meaningless.

SIME CONVENTIONS IS USING OPERATORS :

(i) $$\left(A + B \right)\psi = A\psi + B\psi = B\psi + A\psi = \left(B + A \right)\psi$$

The result of operation by a sum of two operators, on a functin is same as the sum of the results of operations of individual operators on the function. Also, the addition of two

operators is commulative i.e. order of addition is immaterial : $A + B = B + A$

(ii) Although addition (and subtraction) of operators is commutative, the product of two operators may not be commutative always. $AB \neq BA$ and the result depends on the order in which the operators operate. The product of two operation is defined as

$$A\,B\psi = A\;B\psi = A\psi' = \psi''$$

In the operation by AB, it is the operator B that operates first and the result is next operated upon by A to give the final result. In a product, the operators are assumed to operate successilvely, beginning with the operator immediately on the left of the function i.e. beginning with the operator on the extreme right in the operator product we gradually work leftward. This sequence in operations is important since as said earlier, $AB \neq BA$ i.e. multiplication is not commulative.

Example :   Let A be x (multiplication by x)

and B be d/dx (taking first dervative with respect x)

consider a function, $f(x) = x^2$

$AB f(x) = x$ $\$ \dfrac{d}{dx}$ x^2 $\dfrac{d}{dx}$ x^2 $= x(2x) = 2x$ 2

$BA f(x) = x \cdot x =$ $\dfrac{d}{dx}$ x^2 $\dfrac{d}{dx}$ x^2 $\dfrac{d}{dx} = x^3$ $= 3x$ 2

$$\therefore\; AB f(x) - BA f(x) = (AB - BA)f(x) = -x^2 = -f(x)$$

Lifting the fuction,

$$(AB - BA) = -1$$

$(AB - BA)$ is called the commutator of the operators A and B and is represented

as A, B . The value of the commutator is different for different pairs of operators.

If A, B $= 0$ i.e. $AB = BA$ the operators A and B are said to commute with each other and the order of operations is unimportant.

Note that the commutator of two operators is itself an operator. If the commutator is zero. (*a* null operator), it will annihilate its operand. Note also the deep physical meaning of the commutation, behaviour of operators. if two operators commute then the variable corresponding to them can be simultaneously determined with complete precision.

(Does x commute with Px)

(iii) LINEAR OPERATOR

An operator A is said to be linear if the result of operation of A on a sum of functions is same as the results of operation of A on individual functions.

i.e., $A\left[f(x)+g(x)\right]=Af(x)+Ag(x)$

or more generally

$$A\left[af(x)+bg(x)\right]=aAf(x)+bAg(x)$$

(a, b are constants)

Note that not all operators are linear. For example, d/dx is linear but $(\ )^2$ or $\sqrt{\ }$ is not linear.

(iv) HERMITIAN OPERATOR

All operators that we use in quantum mechanics are Hemitian. The Hermiticity of an operator assures real values of the associated variable. The definition of a Hermitian operator is a little difficult to grasp. An operator A is said to be Hemitian if, for any well behaved functions.

ψ_m and ψ_n

$$\int_{all\ space}\psi_m\,A\,\psi_n\,d\tau=\int_{all\ space}\psi_n\,A*\psi*_m\qquad d\tau=\int_{all\ space}\psi^{*}{}_{n}\,A\,\psi_m\,d\tau$$

using Dirac bracket notation,

$$\langle m|A|n \rangle = \langle n|A|m \rangle^*$$

(Only the subscripts are used to designate the functions)

or $\quad A_{mn} = A^*_{nm}$

(v) EIGENVALUE EQUATION

When an operator operates on a function, the function in general, changes into a new function (definition of an operator). However, it is interesting to observe that between certain operators and operands a special relationship exists such that

$$A\,\psi = a\psi \quad \text{(a is a constant)}$$

That is, when the operator operates on the state function the function is not changed except that it gets multiplied by a constant. when this happens, the function is said to be an eigenfunction of the operator and the constant multiplier is said to be the eigenvalue of the operator with respect to that particular function. This interesting relationship embodied in above equation is of great interest in quantum mechanics and is known as the eigenvalue equation.

Exercise : Show that e^{ax} is an eigenfunction of d/dx. What is the eigenvalue ?

$$\frac{d}{dx}\left(e^{ax}\right) = a\,e^{ax} \quad \text{(a = constant)}$$

$\therefore \quad e^{ax}$ is an eigenfunction of $\dfrac{d}{dx}$ belonging to the eigenvalue a.

Exercise : Find out if e^{ax_2} is an eigenfunction of $\dfrac{d}{dx}$.

$$\frac{d}{dx}\left(e^{ax^2}\right) = a.\,e^{ax^2}\,.\,2x = 2\,a\,x.\,e^{ax_2}$$

$$\neq \text{constant } x.\,e^{ax^2}$$

$\therefore e^{ax_2}$ is not an eigenfunction of $\dfrac{d}{dx}$.

The eigenvalue equation has a deep significance. It is the counterpart of experimental

measurement in classical theory. If $A\psi = a\psi$, then the property A associated with the

operator A is well-defined, i.e., known with complete certainty. Its value is a in the particular state ψ. The eigenvalue thus represents the precise result of determination of the variable associated with the operator in the state described by the eigenfunction ψ (eigenstate ψ).

(vii) IMPORTANT THEOREMS ON HERMITIAN OPERATION

THEORAM-1 : *The eigenvalues of Hermitian operators are real.*

***Proof* :** Let ψ be an eigenfunction of an operator A with an eigenvalue a.

$$\therefore A\,\psi = a\psi \qquad\qquad ------(1)$$

Taking the complex conjugate of (1)

$$A\,\psi^* = a^*\,\psi^* \qquad\qquad ------(2)$$

Multiply (1) on the left by $\psi*$ and intergrate

$$\therefore \int \psi*A\psi\ d\tau = a\int \psi*\psi\ d\tau \qquad\qquad ------(3)$$

Similarly multiply (2) on the left by ψ and integrate

$$\therefore \int \psi A\,\psi*\,d\tau = a^*\int \psi\,\psi*\,d\tau \qquad\qquad ------(4)$$

If operator A is hermitian, the left hand sides of equation (3) and (4) must be equal (definition)

$$\int \psi*A\psi\ d\tau = \int \psi A\,\psi*\,d\tau$$

$\therefore$ The right hand sides of (3) and (4) must also be equal, i.e.

$$a\int \psi*\psi\ d\tau = a^*\int \psi\,\psi*\,d\tau$$

$\therefore a = a^*$ 　　　　(Note : $\psi*\psi = \psi\,\psi*$)

This can be true only if a is real.

THEOREM - 2 : Eigenfunctions blonging to different eigenvalues of a Hermitian operator are orthogonal.

Before we can tak up the proof of this theorem, we must understand what are

"orthogonal" functions. So far we have said that an acceptable ψ must be normalized to unity.

i.e,
$$\int \psi * \psi \ d\tau = 1$$

ORTHOGONAL FUNCTIONS :

If $\varphi_1(x)$ and $\varphi_2(x)$ have the property that

$$\int_a^b \varphi^*_2(x) \ \varphi_1(x) \ dx = 0$$

for a certain integral (a, b) then $\varphi_1(x)$ and $\varphi_2(x)$ are said to be "orthogonal" in this interval (a, b).

If we have a whole set of functions $\{\varphi_i\}(x)$ i.e.

$\varphi_1(x)$, $\varphi_2(x)$, ... such that for any pair of functions $\varphi_i(x) \ and \ \varphi_j(x) \ (i \neq j)$

$$\int_a^b \varphi^*_i(x) \ \varphi_j(x) \ dx = 0 \qquad (for \ i \neq j)$$

i.e. such that every pair of functions in the set is orthogonal in the interval (a, b), then the set is said to be an orthogonal set for (a, b). If, in addition, the functions $\varphi_i(x)$ are normalized individually, i.e.

$$\int_a^b \varphi^*_i(x) \ \varphi_i(x) \ dx = 1 \qquad (for \ all \ i)$$

the set is called an orthonormality set for (a,b). Thus the condition for individually

orthonormality and pair-wise orthogonal) is

$$\int_a^b \varphi^*_i(x)\,\varphi_j(x)\,dx = \delta_{ij}$$

δ_{ij} is 'cronecker delta' with the properties that

$$\delta_{ij} = 1 \quad \text{if} \quad i = j$$
$$= 0 \quad \text{if} \quad i \neq j$$

The advantage of having a complete orthonormal set of functions $\{\varphi_i\}$ for a system

is that any arbitrary state ψ of the system can always be completely expressed as a linear combination of the members of the complete orthonormal set.

$$\psi = c_1\varphi_1 + c_2\varphi_2 + \ldots\ldots\ldots = \sum_i c_i \qquad \text{------- (1)}$$

$\varphi_i = \varphi_i(x)$ and c_1 are constants

Multiply (1) by $\varphi^*_n(x)$ and integrate

$$\int \varphi^*_n \psi\,dx = c_1 \int \varphi^*_n \varphi_1\,dx + c_2 \int \varphi^*_n \varphi_2\,dx + \ldots\ldots\ldots + c_n \int \varphi^*_n \varphi_n\,dx + \ldots \qquad \text{------- (2)}$$

All integrals on the right hand side of (2) vanish except

$$\therefore \quad c_n \int \varphi^*_n \varphi_n\,dx = c_n \qquad (\text{Q}\,\varphi_n \; normalized)$$

$$\therefore \quad c_n \int \varphi^*_n \psi\,dx \qquad \text{------- (3)}$$

In the same way all c_i 's can be determined if the expansion is valid.

If φ_i are chosen to be eigenfunctions of F, i.e.

$$F \varphi i = f_i \varphi_i \qquad \text{------- (4)}$$

then $\quad F \psi = c_1 F \varphi_1 + c_2 F \varphi_2 + \dots \dots \dots \qquad \text{------- (5)}$

Multiply (5) on the left by

$$\psi^* = c_1^* \varphi_1^* + c_2^* \varphi_2^* + \dots \dots \dots$$

and integrate

$$\therefore \int \psi^* F \psi \, dx = \int (c_1^* \varphi_1^* + c_2^* \varphi_2^* + \dots \dots \dots) F (c_1 \varphi_1 + c_2 \varphi_2 + \dots \dots) \, dx$$

$$= c_1^* c_1 \int \varphi^* F \varphi_1 \, dx + c_1^* c_2 \int \varphi_1^* F \varphi_2 \, dx + \dots \dots \qquad \text{------- (6)}$$

All terms of the type $a_m^* a_n \int \varphi_m^* F \varphi_n \, dx$ on the right hand side of (6) must vanish due to the orthogonality of φ_m and φ_n (m ≠ n). The only terms remaining are those with m = n, e.g.

$$c_n^* c_n \int \varphi_n F \varphi_n \, dx = c_n^* c_n f_n \int \varphi_n^* \varphi_n \, dx$$

$$= |c_n|^2 f_n \qquad \text{------- (7)}$$

$$\therefore \int \psi^* F \psi \, dx = \overset{F}{} = F = |c|_1^2 f_1 + |c|_2^2 f_2 + \dots \dots \qquad \text{------- (8)}$$

In (8), $|c_i|^2 = P_i$ is the probability or weight factor of the eigenvalue f_i.

$$F = \underset{i}{\sum_i} P f_i \qquad \text{------- (9)}$$

Thus the expectation value of F (i.e. the mean value of F) is the sum of the probability - weighted eigenvalues f_i.

Proof of Theorem-2 :

Let ψ_1 and ψ_2 be the eigenfunctions of the Hermitian operator A belonging to the eigenvalues a_1 and a_2. respectively.

$$A\psi_1 = a_1 \psi_1 \qquad\qquad\qquad \text{------ (1)}$$

$$A\psi_2 = a_2 \psi_2 \qquad\qquad\qquad \text{------ (2)}$$

From (1) : $\quad \int \psi_2{}^* A\psi_1 d\tau = a_1 \int \psi_2{}^* \psi_1 \, d\tau \qquad\qquad \text{------ (3)}$

Since the operator A is Hermitian,

$$\int \psi_2{}^* A\psi_1 \, d\tau = \int \psi_1 \, A \, \psi_2{}^* \, d\tau$$

$$= a_2 \int \psi_1 \psi_2{}^* \, dt \qquad\qquad (a_2{}^* = a_2) \qquad\qquad \text{------ (4)}$$

$$\therefore a_1 \int \psi_2{}^* \psi_1 \, d\tau = a_2 \int \psi_2{}^* \psi_1 \, d\tau$$

$$\therefore (a_1 - a_2) \int \psi_2{}^* \psi_1 \, d\tau = 0 \qquad\qquad \text{------ (5)}$$

If $a_1 \neq a_2$ (i.e. ψ_1 and ψ_2 belong to different eigenvalues i.e. are non-degenerate), then

$$\int \psi_2{}^* \psi_1 \, d\tau = 0$$

ψ_1 and ψ_2 must be orthogonal.

Schmidt orthogonalization :

If $a_1 = a_2$ in (5) (i.e. ψ_1 and ψ_2 belong to the same eigenvalue, i.e., are degenerate) then since $a_1 - a_2 = 0$, $\int \psi_2{}^* \psi_1 \, d\tau$ may or may not be zero i.e. ψ_1 and ψ_2 are not necessarily orthogonal. However a set of orthogonal fuctions can always be found.

$$\text{Let } a_1 = a_2 = a$$

$$\therefore A\psi_1 = a\psi_1$$

$$A\psi_2 = a\psi_2$$

Let $\quad \int \psi_2{}^* \psi_1 \, d\tau = b \neq 0 \qquad\qquad \text{------ (6)}$

consider $\qquad \psi_2' = \psi_2 - b\psi_1$ $\qquad\qquad\qquad\qquad$ ------- (7)

$\therefore \displaystyle\int \psi_2^* \psi_1 \, d\tau = \int (\psi_2 - b\psi_1)^* \psi_1 \, d\tau$

$\qquad\qquad\qquad = \displaystyle\int \psi_2^* \psi_1 \, d\tau - b \int \psi_1^* \psi_1 \, d\tau$

$\qquad\qquad\qquad = b - b$

$\qquad\qquad\qquad = 0$

Thus ψ_1 and ψ_2^* are orthogonal.

$A\psi_2' = A(\psi_2 - b\psi_1) = A\psi_2 - b A\psi_1$

$\qquad\qquad\qquad = a\psi_2 - ab\psi_1$

$\qquad\qquad\qquad = a(\psi_2 - b\psi_1) \qquad = a\psi_2'$

Thus the linear combination ψ_2' is an eigenfunction of A and belongs to the same eigenvalue a as does ψ_2.

(ix) HOW TO CONSTRUCT OPERATORS ?

An operator is the quantum mechanical device of meauring a variable and so we need to know the operator appropriate to a given variable. These operators are constructed according to the following prescription :

Step : 1 Begin with the classical mathematical expression for the variable of interest and express it in terms of coordinates and momenta of the system.

Step : 2   Make the following replacement :

a.   Leave time and all other coordinates as they are.

$$x \rightarrow x \qquad \text{(operation is multiplication by x)}$$

$$q \rightarrow q \qquad \text{(multiplication by q)}$$

b.   Replace all components of momenta, pq_i by

$$i\mathsf{H} \ \frac{\partial}{\partial q_i} \equiv \frac{\mathsf{H}}{i} \frac{\partial}{\partial q_i} \qquad\qquad \left(\mathsf{H} = h/2\pi \right)$$

Note that (*i*) if the variable is some function of coordinates and momenta, then. the operator is the same function of the corresponding operators. e.g. the z-component of angular momentum is defined as

$$Lz = x\,P_y \ - y\,P_x$$

$\therefore$ The corresponding operator will be

$$\$_{Lz=x} \ \$ \frac{\mathsf{H}}{\iota} \frac{\partial}{\partial y} \quad -y \ \$\_ \frac{\mathsf{H}}{i} \frac{\partial}{\partial x}$$

$$\cdots{}_{\mathsf{-iH}} \quad \$_x \frac{\partial}{\partial y} - \$_y \frac{\partial}{\partial x}$$

(ii)   If the function is such that the order of factors is important, then, the operator must be constructed so that it is Hermitian.

Examples : 1 Kinetic energy of a particle (T) :

$$T = \frac{1}{2} mv^2 = \frac{(mv)^2}{2m} = \frac{p^2}{2m} = \frac{1}{2m}\left(Px^2 + Py^2 + Pz^2\right)$$

$$\therefore\, T = \frac{1}{2m}\left[-i\hbar\frac{\partial}{\partial x} \cdot -i\hbar\frac{\partial}{\partial x} + -i\hbar\frac{\partial}{\partial y} \cdot -i\hbar\frac{\partial}{\partial y} + -i\hbar\frac{\partial}{\partial z} \cdot -i\hbar\frac{\partial}{\partial z}\right]$$

$$= -\frac{\hbar^2}{2m}\left[\frac{\partial^2}{\partial x^2} + \frac{\partial^2}{\partial y^2} + \frac{\partial^2}{\partial z^2}\right]$$

$$= -\frac{\hbar^2}{2m}\nabla^2$$

∇^2 is called the Laplacian operator.

$$\nabla^2 = \frac{\partial^2}{\partial x^2} + \frac{\partial^2}{\partial y^2} + \frac{\partial^2}{\partial z^2}$$

TOTAL ENERGY (E) OF A SYSTEM

Total energy is the single most important peoperty of a system in which we shall be interested. The corresponding operator is called the Hamiltonian operator or just the Hamiltonian, H.

Sinlge particle system :

$E = T + V$ (V is the potential energy)

$$\therefore E = \frac{1}{2m}\left(Px^2 + Py^2 + Pz^2\right) + V(xyz)$$

This is the classical Hamilton's function from which the corresponding operator derives its name.

$$\therefore H = T + V = -\frac{\hbar^2}{2m}\nabla^2 + V(x,y,z)$$

Two particles (identical) system : $(m_1 = m_2 = m)$

$$E = \frac{1}{2}mv_1^2 + \frac{1}{2}mv_2^2 + V \quad = P^2/2m + P^2/2m + V$$

$$\therefore H = - \frac{H^2}{2m} \left(\nabla_1^2 + \nabla_2^2 \right) + V (x_1\, y_1\, z_1\, x_2\, y_2\, z_2)$$

(Note) V here is the potential energy of the whole system) If the two particles are not indentical $(m_1 \neq m_2)$:

$$H = - \frac{H^2}{2} \frac{1}{m_1} \nabla_1^2 + \frac{1}{m_2} \nabla_2^2 + V$$

We shall be generally concerned wih electronic Hamiltonians for atomic and molecular systems where all particles of interest are electrons (indentical).

N-electron system :

$$E = \sum_i T_i + V$$

$$\therefore H = - \frac{H^2}{2m} \sum_i \nabla_i^2 + V (q_i) \ldots\ldots\ldots\ldots \qquad q = x\,,\, y\,,\, z \quad i = 1,\ldots\ldots N$$

Hydrogen atom : (z = 1)

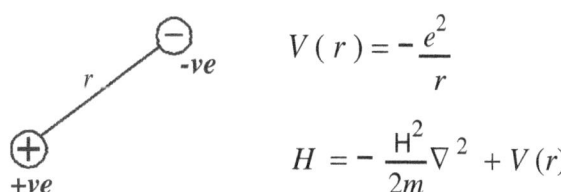

$$V(r) = - \frac{e^2}{r}$$

$$H = - \frac{H^2}{2m} \nabla^2 + V(r)$$

Helium atom : (z = 2)

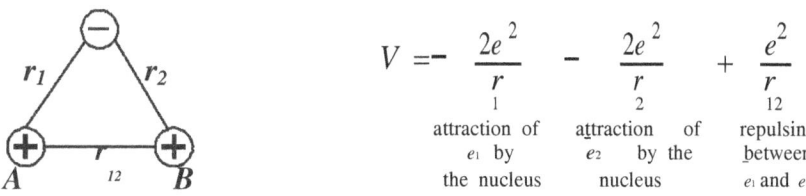

$$V = - \frac{2e^2}{r_1} - \frac{2e^2}{r_2} + \frac{e^2}{r_{12}}$$

| attraction of e_1 by the nucleus | attraction of e_2 by the nucleus | repulsing between e_1 and e_2 |

$$H = - \frac{H}{2m} \left(\nabla_1{}^2 + \nabla_2{}^2 \right) + V$$

Carbon atom : (z = 6)

There are six e outside the nucleus of charge +6e. The total P.E. (v) comprises six

attractive terms of the type $\dfrac{-6e_2}{\overline{\quad}}$, one for each electron and fifteen interelectronic

$$\overline{\text{repulsive } r_i}$$

terms of the type $+\dfrac{e^2}{r_{ij}}$, one for each distinct repelling pair of electrons.

Thus :

$$V = - \frac{6e^2}{r_1} - \frac{6e^2}{r_2} - \frac{6e^2}{r_3} - \frac{6e^2}{r_4} - \frac{6e^2}{r_5} - \frac{6e^2}{r}$$

$$+\frac{e^2}{r_{12}} + \frac{e^2}{r_{13}} + \frac{e^2}{r_{14}} + \frac{e^2}{r_{15}} + \frac{e^2}{r_{16}} \qquad \leftarrow \text{repulsion of } \overline{e_1} \text{ by all other } \overline{e}\text{'s}$$

$$+\frac{e^2}{r_{23}} + \frac{e^2}{r_{24}} + \frac{e^2}{r_{25}} + \frac{e^2}{r_{26}} \qquad \leftarrow \text{repulsion of } \overline{e_2} \text{ by all other } \overline{e}\text{'s}$$

$$+\frac{e^2}{r_{34}} + \frac{e^2}{r_{35}} + \frac{e^2}{r_{36}} \qquad \leftarrow \text{repulsion of } \overline{e_3} \text{ by all other } \overline{e}\text{'s}$$

$$\frac{e^2}{r_{45}} + \frac{e^2}{r_{46}} \qquad \leftarrow \text{repulsion of } \overline{e_4} \text{ by all other } \overline{e}\text{'s}$$

$$\frac{e^2}{r_{56}} \qquad \leftarrow \text{repulsion of } \overline{e_5} \text{ by all other } \overline{e}\text{'s}$$

H 2 6   6 1   6 1

$$H = -\frac{\mathrm{H}}{2m}\sum_{i=1}\nabla_i^2 - 6e^2\sum_{i=1}\frac{1}{r_i} + e^2\sum_{i<j}\frac{1}{r_{ij}}$$

In general for an atom with atomic number Z,

$$H = -\frac{\mathrm{H}^2}{2m}\sum_{i=1}^{z}\nabla_i^2 - Ze^2\sum_{i=1}^{z}\frac{1}{r_i} + e^2\sum_{i<j}^{z}\frac{1}{r_{ij}}$$

Exercise : State explicitly the potential function and hence write the Hamiltonian operators for (i) Lithium atom (z = 3) (ii) Oxygen atom (z = 8) and (iii) Hydrogen molecule, H_2.

THIRD POSTULATES OF QUANTUM MECHANICS

SCHRODINGER EQUATION:

We have said earlier that ψ is a rich source of information about the state and have described how operators can serve to extract information contained in ψ. So far we have said nothing about the source of this powerful quantity: How to get this ψ? This question is answered in Postulate 3 which states:

The possible state functions ψ of a system are the solution of the differential equation.

$$H\psi = i\mathrm{H}\ \frac{\partial\psi}{\partial t}$$

This equation, known as the time-dependent Schrodinger equation, is the fundamental equation of motion in quantum mechanics. It describes how ψ develops with time. It must be mentioned here that, although it is easy to write the Schrodinger equation for any system, it is very difficult to solve it exactly for most of the systems. In fact it has been solved exactly for only the one-electron systems (hydrogen atom and hydrogen-like ions). For all other systems. only the approximate solutions are obtained using standard approximation methods such as the variation principle and the perturbation methods.

Let us practice writing Schrodinger equation using simple cases.

Chapter-3

QUNTUM MECHANICS OF SIMPLE SYSTEM WITH THE CONSTANT POTENTIAL ENERGY :

QUANTUM MECHANICS OF SIMPLE SYSTEMS
PARTICLES IN A BOX PROBLEMS

We shall now apply the principles discussed above to some simple artificial systems The most obvious virtue of these elementary applications of Schrodinger equation is their simplicity such that without bringing in excessive mathematics it is possible to obtain exact solutions. At the same time, these applications very elagently illustrate many of the fundamental aspects such as quantization, degeneracy etc. of the quantum theory. Again, the results are of real chemical inerest since these artificial systems serve as convenient models for situations of chemical interest.

FREE PARTICLE :

This is a trivial case but its discussion can lead to better appreciation of particle in a box problem to follow.

A free particle is a particle moving in a constant potential field with no coordinate restriction.

$$V = \text{constant} = 0 \qquad \text{(arbitrarily)}$$

Assume a linear system i.e., $V(x) - 0$. Our interest is to find the wave functions and energy for this free particles.

$$H = -\frac{H^2}{2m}\frac{d^2}{dx^2}$$

$\therefore$ Schrodinger equation will be

$$-\frac{H^2}{2m}\frac{d^2\psi}{dx^2} = E\psi$$

$$\therefore \quad \frac{d^2\psi}{dx^2} = -\frac{2mE}{\;}\psi$$

$$\therefore \quad \frac{d^2 \psi}{dx^2} + k^2 \psi = 0 \qquad \text{Where} \qquad k^2 = \frac{2m E}{H^2} = \frac{Px^2}{H}$$

We can easily guess the solutions which could be

$\psi (x) = A$
$\sin kx \ \psi (x$
$) = A \cos kx$
$\psi (x) = A$
$. e^{\pm ikx}$

(i)   Note that there is no restriction on k and hence on E, i.e. the energy of a free particle is not quantized.

(ii) $\psi^* \psi = A^* A \ e^{-ikx} . e^{ikx} = A^* A =$ constant so that

all positions are equally probable for a free particle i.e. Position is uncertain.

(iii)   Let us obtain Px.

$$P_x \psi = \frac{H}{i} \frac{d}{dx} \left(A e^{ikx} \right)$$

$$= H k . A e^{ikx} = \left(H k \right) \psi$$

$\therefore \psi$ us an eigenfunction of Px with the eigenvalue hk.

$\therefore$ exact value of $Px = H k = \sqrt{2m E}$ which is the classical result.

PARTICLE IN A ONE-DIMENSIONAL (LINEAR) BOX

Since the position of a free particle is completely uncertain let us confine it between

$x = 0$ and $x = L_x$ so that we know for certain that the particle exists somewhere between these limits. This problem is called the "particle in the box." It corresponds to the potential function.

$V(x) = \infty$	$V(x) = const = 0$	$V(x) = \infty$	$\infty \quad x \leq 0$
for $x \leq 0$	$0 < x \leq Lx$	for $x \geq Lx$	$V(x) = 0 \quad 0 < x < Lx$
$\psi(x) = 0$	$\psi(x)$ finite	$\psi(x) = 0$	$\infty \quad x \geq Lx$

So that the particle is constrained within two infinite potential barriers.

For the infinite potential region outside the box ($x < 0$ and $x > Lx$) it can be shown that the only acceptable solution is $\psi(x) = 0$ i.e. the particle cannot get out of the box.

For the inside region ($0 < x < Lx$), $V(x) =$ and the Schrodinger equation is same as that for a free particle

$$\frac{d^2\psi}{dx^2} + k^2\psi = 0 \qquad \text{where } k^2 \quad = \frac{2mE}{H^2} = \frac{Px}{H}^2$$

The solution $\psi(x)$ must be "well-behaved". Since $\psi(x) = 0$ everywhere outside, continuity will require that $\psi(x)$ vanishes at the walls (x $= 0$ and x $= L_x$) Thus the boundary conditions on acceptable solutions are :

$$\psi(0) = 0$$
$$\psi(Lx) = 0 \qquad \qquad \text{------(1)}$$

i.e. $\psi(x) = 0$ at $x \leq 0$ and $x \geq Lx$

Equatio (1) has a general solution

$$\psi(x) = A \sin kx + B \cos kx \qquad \text{------(2)}$$

Applying the boundary conditions,

$$\psi(0) = A \sin 0 + B \cos 0 = B = 0$$
$$\therefore \psi(x) = A \sin kx$$

Simularly, $\psi(Lx) = A \sin k Lx = 0$

$$\therefore A \neq 0, \quad k Lx = n_x \pi \qquad \text{------(3)}$$

$$\text{Where} \quad n_x = 1, 2, 3, \dots\dots\dots$$

$$\therefore k = \frac{n_x \pi}{\cdot}$$

$$\therefore \psi_{n_x}(x) = A \sin \frac{n_x \pi_x}{Lx} \qquad \text{------(4)}$$

EVALUATION A:

The constant A in (4) can be obtained by applying the normalzation condition.

$$\int_0^{Lx} \psi_{nx}^*(x)\, \psi_{nx}(x)\, dx = 1$$

$$\therefore \int_0^{Lx} A \sin \frac{n_x \pi_x}{Lx} \cdot A \sin \frac{n_x \pi_x}{Lx}\, dx = 1$$

$$\text{Use} \qquad \sin^2 \alpha = \frac{1}{2}(1 - \cos 2\alpha)$$

$$\therefore A^2 \int_0^{Lx} \sin^2 \frac{n_x \pi_x}{Lx}\, dx = 1$$

$$\therefore \frac{A^2}{2} \int_0^{Lx} dx - \int_0^{Lx} \cos \frac{2n_x \pi_x}{Lx}\, dx = 1$$

$$\therefore \frac{A^2}{2} \left. 1 \right|_1^{Lx} - 0 = 1$$

$$\therefore \frac{A^2}{2}[Lx] = 1$$

$$\therefore A^2 = \frac{2}{Lx}$$

$$\therefore A = \pm \sqrt{\frac{2}{Lx}} \qquad (n_x = 1, 2, 3, \dots\dots\dots\dots) \qquad \text{------(5)}$$

ZERO POINT ENERGY (ZPE) :

There is no energy state corresponding to the particle at rest with KE = 0. Why ?

for E - 0, $\psi = 0$ everywhere in the box. The box is empty !

More convincing argument derives from the uncertainty principle. The best that we can say is that the particle is located some where in the box $0 < x < Lx$.

$$\therefore x = Lx$$

If $E = 0$, $Px = 0$ and $Px = 0$.

$$x \ . \ Px = 0$$

This violetes the uncertainty principle.

The lowest energy corresponds to the lowest value of n_x $(= 1)$ and is

$$E_1 = h^2 / 8m L_x^2 = E_0$$

called the zero point energy (ZPE). Zero point energies are characteristic of bound systems.

— —

PARTICLE IN THREE DIMENSIONAL CUBIC BOX

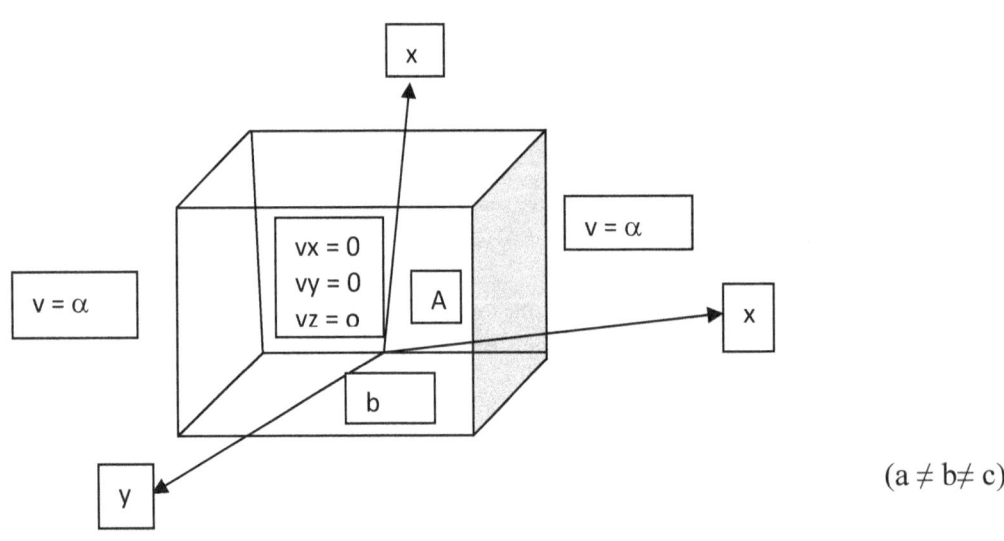

$(a \neq b \neq c)$

The schrodinger equation for a constant- energy in a field- free cubic box of side a in which V is zero for any value of x, y, or z between zero and a but infinite outside these limits, This is

$$\frac{d^2\Psi}{dx^2} + \frac{d^2\Psi}{dy^2} + \frac{d^2\Psi}{dz^2} + \frac{8\pi^2 m}{h^2} (E-V) \ \Psi = 0 \ \ldots\ldots\ldots\ldots (I)$$

$$\frac{d^2\Psi}{dx^2} + \frac{d^2\Psi}{dy^2} + \frac{d^2\Psi}{dz^2} + \frac{8\pi^2 m}{h^2} E \ \Psi = 0 \ \ldots\ldots\ldots\ldots\ldots (II) \quad (\because v = 0)$$

When Ψ is a function of three independent variable x, y and z. It can be shown to be a product of three function separately dependent on the same variable. i.e. (તરંગ વિધેય Ψ ને ત્રણે દિશામાં ત્રણ સ્વતંત્ર તરંગવિધેયોના ગુણાકાર સ્વરૂપે દર્શાવી શકાય.)

$$\Psi (x, y, z) = \Psi x \ \Psi y \ \Psi z \ \ldots\ldots\ldots\ldots (III)$$

$\Psi(x)$ is the function of x and independent y & z, $\Psi(y)$ is the function of y and independent from x and y.

i.e.
$$\left. \begin{array}{l} \dfrac{d^2\Psi}{dx^2} = \Psi(y) \ \Psi(z) \dfrac{d^2\Psi(x)}{dx^2} \\[2mm] \dfrac{d^2\Psi}{dy^2} = \Psi(x) \ \Psi(z) \dfrac{d^2\Psi(y)}{dy^2} \\[2mm] \dfrac{d^2\Psi}{dz^2} = \Psi(x) \ \Psi(y) \dfrac{d^2\Psi(z)}{dz^2} \end{array} \right\} \quad \boxed{\text{iv}}$$

put the value of (III) & (IV) in Eq. (II)

$$\Psi(y) \ \Psi(z) \frac{d^2\Psi(x)}{dx^2} + \Psi(x) \ \Psi(z) \frac{d^2\Psi(y)}{dy^2} + \Psi(x) \ \Psi(y) \frac{d^2\Psi(z)}{dz^2} + \frac{8\pi^2 mE}{h^2} \Psi(x) \ \Psi(y) \ \Psi(z)$$

$$\ldots\ldots\ldots\ldots (V)$$

Divide by $\Psi(x), \Psi(y), \& \Psi(z)$

Four steps in Eq. (vi) 4th step is constant if change the value of 1st step by keeping constant the 2nd & 3rd steps.

i.e.
$$\frac{1}{\Psi_x} \times \frac{d^2\Psi(x)}{dx^2} + \frac{8\pi^2\,mE_x}{h^2} = 0$$
$$\frac{1}{\Psi_y} \times \frac{d^2\Psi(y)}{dy^2} + \frac{8\pi^2\,mE_y}{h^2} = 0$$
$$\frac{1}{\Psi_z} \times \frac{d^2\Psi(z)}{dz^2} + \frac{8\pi^2\,mE_z}{h^2} = 0$$

$\left. \phantom{\begin{array}{c}1\\1\\1\\1\end{array}} \right\}$ ———— VII

Multiply 1st eq by $\Psi(x)$, 2nd by $\Psi(y)$ and 3rd by $\Psi(z)$ in Eq. – VII

$$\frac{d^2\Psi(x)}{dx^2} + \frac{8\pi^2\,mE_x}{h^2}\,\Psi(x) = 0$$
$$\frac{d^2\Psi(y)}{dy^2} + \frac{8\pi^2\,mE_y}{h^2}\,\Psi(y) = 0$$
$$\frac{d^2\Psi(z)}{dz^2} + \frac{8\pi^2\,mE_z}{h^2}\,\Psi(z) = 0$$

———— VII

Eq- VIII are the same like particle in dimensional box, therefore the normalized wave functions and the Equ. of energy are

(i) Normalized wave functions

$$\Psi(x) = \sqrt{\frac{2}{a}}\,\sin\frac{n\pi}{a}x$$

$$\Psi(y) = \sqrt{\frac{2}{b}}\,\sin\frac{n\pi}{b}y$$

$$\Psi(z) = \sqrt{\frac{2}{c}}\,\sin\frac{n\pi}{c}z \qquad \ldots\ldots\ldots\ldots\ldots\ (IX)$$

$$\Psi\,(x,y,z) = \sqrt{\frac{8}{abc}}\,\sin\frac{n\pi}{a}x.\ \sin\frac{n\pi}{b}y.\ \sin\frac{n\pi}{c}z$$

(ii) Energy : $E = E_x + E_y + E_z$

$$\therefore E = \frac{n^2_x h^2}{8ma^2} + \frac{n^2_y h^2}{8mb^2} + \frac{n^2_z h^2}{8mc^2}$$

$$\therefore E = \frac{h^2}{8ma^2} [n^2x + n^2y + n^2z]$$

DEGERENCY

Energy Equation for three dimensional cubic box is

$$E = \frac{h^2}{8m} [\frac{n^2_x}{a^2} + \frac{n^2_y}{b^2} + \frac{n^2_z}{c^2}] ----------------- (I)$$

For cubic box (સમઘન) a = b = c

$$\therefore E = \frac{h^2}{8ma^2} [n^2x + n^2y + n^2z] --------------(II)$$

Put the value n= 0 in Eq- (2) i.e. $n_x = n_y = n_z = 0$ there for E = 0 put the value n=0 in Normalization wave function. So Ψ become 0 ($\Psi^2 = 0$) means probality to finding the $\bar{e}$ in box is zero i.e. There is no particle in box. But particle is in box it is real fact. So value of ($\Psi^2 = 0$) is not possible & also n≠ 0 (n = 0 હોઈ શકે નહી)

So, minimum energy level (નીચી કવોન્ટમ અવસ્થા) $n_x = n_y = n_z = 1$ put the value of this sets of Quantum Numbers in Eq = II

$$E = \frac{h^2}{8m} (1^2 + 1^2+ 1^2)$$

$$\therefore E = \frac{3 h^2}{8ma^2}$$

Here, for this energy level, only one set of Quantum number (1,1,1). So we said this energy level is non- degenerate energy level.

Then for higher energy level, we get the sets of quantum numbers are

(1) $n_x =2$, $n_y =1$, $n_z = 1$
(2) $n_x =1$, $n_y =2$, $n_z = 1$
(3) $n_x =1$, $n_y =1$, $n_z = 2$

For this energy levels

$$\therefore E = \frac{6 h^2}{8ma^2}$$

After this position, Also we get three sets of Quantum numbers, which have the same energy. (ત્રણેય કવોન્ટમ અંકોની શક્તિ સમાન હોય છે.) and this energy energy levels are try-deqenerate energy levels.

Then higher energy levels are as under.

$E_5 = \frac{12\,h^2}{8ma^2}$ | _____ (2,2,2) _____ Non – degenerate

$E_4 = \frac{11\,h^2}{8ma^2}$ | (3,1,1)   (1,3,1)   (1,1,3) _____ Try – degenerate

$E_3 = \frac{9\,h^2}{8ma^2}$ | (2,2,1)   (2,1,2)   (1,2,2) _____ Try – degenerate

$E_2 = \frac{6\,h^2}{8ma^2}$ | (2,1,1)   (1,2,1)   (1,1,2) _____ Try – degenerate

$E_1 = \frac{3\,h^2}{8ma^2}$ | _____ (1,1,1) _____ Try – degenerate

           | Zero Point Energy (ZPE) $\updownarrow$

$\leftarrow$ Quantum numbers (Energy) $\rightarrow$

Degeneracy (સમશક્તિપણુ) of energy levels is depend on the symmetry of cubic box.   (a=b=c). if (a $\neq$ b$\neq$ c) decrease the degeneracy and increased the energy levels.

One Dimensional Box

Put the value of K in Eq – (6)

Ψ = Asin ($\frac{n\pi}{L}$) x -------------- (7)

Eq- (7) shows the wave function of $\bar{e}$ in one dimensional box

Energy :- $K^2 = \frac{8\pi^2\,mE}{h^2}$

     ($\frac{n\pi}{L}$)$^2 = \frac{8\pi^2\,mE}{h^2}$ (where K = $\frac{n\pi}{L}$)

     $\therefore \frac{n^2\pi^2}{L^2} = \frac{8\pi^2\,mE}{h^2}$

$\therefore E = \dfrac{n^2 \pi^2}{L^2} \times \dfrac{h^2}{8\pi^2 mE}$

$\therefore E = \dfrac{n^2 h^2}{8mL^2}$ (Energy Equation for the particle in dimensional box)

CONSERVATIVE SYSTEM :

By a conservative system we mean any isolated system, not acted upon by external forces (V depends only on position coordinates) and with no internal dissipative forces, whose total energy (E = T + V) remains constant with time. Such a system exists in quantized stationary states which are described by state functions of a special form. ψ

$(x, t) = \psi (x) . \varphi (t)$

Space
Time
part part

Which permits the SEPARATION OF VARIABLES so that the Schrodinger equation can be split into two independent equations.

$$H\,\psi = i\,\mathrm{H}\dfrac{\partial \psi}{\partial t}$$ -----(1)

Assume $\psi (q, t) = \psi (q).\varphi (t)$ -----(2)

$\therefore \quad H\,\underset{(\ )}{\psi\varphi} = i\,\mathrm{H}\dfrac{\partial}{\partial t}\,\underset{(\ )}{\psi\varphi}$

$\therefore \quad \varphi H\,\psi = i\,\mathrm{H}\psi\dfrac{\partial \varphi}{\partial t}$

$\therefore \quad \dfrac{H\,\psi}{\psi} = \dfrac{i\mathrm{H}}{\varphi\partial t}.\dfrac{\partial \varphi}{}$ -----(3)

The left hand side of (3) depends on the position or space coordinates q (i.e., x y z) while the right hand side depends only on t. Since x, y, z are independent of t, eq. (3)

can be true only if the two sides separately equal some constant, say E.

$$\therefore \quad \frac{H\,\psi}{\psi} = E \qquad \text{or} \qquad H\,\psi = E\psi \qquad \text{------(4)}$$

and
$$\frac{iH}{\varphi} \cdot \frac{\partial\varphi}{\partial t} = E \qquad \text{------(5)}$$

We see that assumption (1) has enabled us to separate the space and the time parts of the time-dependent Schrodinger equation.

Equation (5) is easily solved. It can be rewritten as

$$\frac{\partial\varphi}{\varphi H} = \underline{iE}\,\partial t$$

Upon integration,
$$\varphi = -\frac{iEt}{H}$$

$$\therefore \varphi(t) = e^{-iEt/H} \qquad \text{------(6)}$$

Equation (4) is the familiar time independent Schrodinger equation :

$H\,\psi = E\psi$. This is an eigenvalue equation for the Hemiltonian operator (corresponding to the total energy) and the eigenvalue E represents the energy of the stationary state (We were foresighted to choose E to represent the separation constant !)

In general : $\qquad H\,\psi_i = E_i\,\psi_i \qquad \text{------(7)}$

Where $\{\psi_i\}$ represents the whole set of eigenstates and $\{E_i\}$ the corresponding energy levels. Each member of $\{\psi_i\}$ corresponds to a particular solution of the form.

$$\psi_i(q_i, t) = \psi_i(q_i) \cdot \exp.\left(-i\,E_i\,t/H\right) \qquad \text{------(8)}$$

with $\quad H\,\psi_i = E_i\,\psi_i \qquad \text{------(9)}$

the allowed eigenstates ψ_i.

$$i(i, t) = i^{(i)} \cdot \exp.\left(-i\,E\,t/H\right) \qquad \text{------(10)}$$

with $\qquad H_i \psi_i = E_i \psi_i$ ------(11)

Consider now the probability of observing the system in any of the allowed eigenstates,

$$\psi_i^* \psi_i = \psi_i^* \cdot e^{iE_i t/H} \cdot \psi_i \, e^{-iE_i t/H}$$
$$= \psi_i^* \psi_i$$

The probability distribution is independent of time. Each eigenstate is unique and rightly are these eigenstates called the stationary states. We will summarize the above discussion and put it as a part of P.3.

ORTHOGONALITY :

consider ψ_n and ψ_m with $n \neq m \, (n > m)$.

$$\int_0^{L_x} \psi_n \psi_m \, dx$$

$$= \frac{2}{Lx} \int_0^{L_x} \sin \frac{n\pi x}{Lx} \sin \frac{m\pi x}{Lx} \, dx$$

Use : $\qquad \sin A \, \sin B = \frac{1}{2} \cos(A - B) - \cos(A + B)$

$$= \frac{2}{Lx} \frac{1}{2} \int_0^{L_x} \cos(n-m) \frac{\pi x}{Lx} \, dx - \int_0^{L_x} \cos(n+m) \frac{\pi x}{Lx} \, dx$$

$$= \frac{1}{Lx} \left[\frac{Lx}{(n-m)\pi} \sin(n-m) \frac{\pi x}{Lx} - \frac{Lx}{(n+m)\pi} \sin(n+m) \frac{\pi x}{Lx} \right]_0^{L_x}$$

$$= 0$$

$\therefore \quad \psi_m$ and ψ_n (m $\neq$ n)are orthogonal.

EIGENVALUES : ENERGY LEVELS

Recall the condition (eq. (4)) obtained on imposing the boundary condition.

$$k\ L_x\ = n_x\ \pi$$

$$\therefore\ \frac{P_x}{H}.L_x\ = n_x\ \pi$$

$$\therefore\ P_x\ = n_x\ .\ \frac{H}{2L_x}$$

We find that only discrete, special values of P_x are permitted ! P_x can be $\dfrac{H}{2L_x}$ or

any integral multiple of it. In other words. P_x and as a consequence the energy of the particle is quantized.

$$E_{n_x}\ = \frac{P^2_x}{2m}$$

$$\therefore\ E_{n_x}\ = n^2_x\ \frac{H^2}{8m\ L^2_x}$$

This is in contrast to the free particle without any constraint. Thus quantization is a consequence of boundary conditions imposed,

Alternative Method for E_{n_x} **:**

This is the usual method of calculating energy corresponding to a state function.

We have $\quad H\ \psi_n = E_n\ \psi_n$

$$\therefore\ \psi^*_n\ H\ \psi_n\ = \psi^*_n\ E\ \psi_n\ = E\psi^*_n\ \psi_n$$

Integrating both sides and rearranging.

$$E_n = \int \psi^*_n\ H\ \psi_n\ d\tau\ /\psi^*_n\ \psi_n\ d\tau$$

$$= \int \psi^*_n\ H\ \psi_n\ d\tau\ \text{ if } \psi_n\ \text{ is normalized}$$

Using (10) with the particle in a box function ψ_{n_x} :

$$E_{n_x} = \frac{2}{L_x} \int_0^{L_x} \sin \frac{n_x \pi x}{L_x} \left[-\frac{H^2}{2m} \frac{d^2}{dx^2} \sin \frac{n_x \pi x}{L_x} \right] dx$$

$$\frac{H^2}{2m} \cdot \frac{n_x^2 \pi^2}{L_x^2} \cdot \sin \frac{n_x \pi x}{L_x}$$

$$= n_x^2 \cdot \frac{\pi^2 H^2}{2mL_x^2} \int_0^{L_x} \frac{2}{L_x} \sin^2 \frac{n_x \pi x}{L_X} dx$$

$$\equiv \int_0^{L_x} \psi_{n_x}^* \, \psi_{n_x} \, dx = 1$$

(v)   Other properties of the system :

(a)   Momentum Px in the ground state, E_1,

The operator P_x is $\frac{H}{i} \frac{d}{dx}$. It can be shown that ψ_1 is not an eigenfunction of P_x.

$$P_x \psi_1 = \frac{H}{i} \frac{d}{dx} \sqrt{\frac{2}{L_x}} \sin \frac{\pi x}{L_x} = \frac{H}{i} \sqrt{\frac{2}{L_x}} \cdot \frac{\pi x}{L_x} \cos \pi \frac{x}{L_x}$$

Therefore, measurement of Px will give a distribution of results. We cannot have exact knowledge of Px. However, we can determine the mean value of Px from

$$\bar{P}_x = P_x = \int_0^{L_x} \psi_1 P_x \psi_1 \, dx \qquad (\psi_1 \text{ is normalized})$$

$$\therefore P_x = \frac{2}{L_x} \int_0^{L_x} \sin \frac{\pi x}{L_x} \frac{H}{i} \frac{d}{dx} \sin \frac{\pi x}{L_x} dx$$

$$= \frac{2}{L_x} \frac{\pi x}{L_x} \frac{H}{i} \int_0^{L_x} \sin \frac{\pi x}{L_x} \cos \frac{\pi x}{L_x} dx$$

—   —   —

Use : $\sin 2\alpha = 2 \sin\alpha \cos\alpha$

$$= \overline{i\,L_x^2} \cdot \overline{2}^1 \int_0^{Lx} \sin \frac{2\pi x}{L_x}\ dx$$

$$= \frac{h}{2i\,L_x^2} - \frac{L}{2\pi} \cos 2\pi \frac{x}{Lx}\Big|_0^{Lx}$$

$$\therefore P \quad \langle\ \rangle \quad \frac{H}{i\,L_x^2} \quad -\frac{L_x}{2\pi} \quad --\frac{L_x}{2\pi}$$

$$\therefore \quad \langle P_x \rangle = 0$$

Consider P_x^2 in the lowest state, ψ_1 . The operator of interest is $-H^2\ d/dx^2$.

$$-H^2 \frac{d^2\psi_1}{dx^2} = H^2 \cdot \frac{\pi^2}{L_x^2}\ \psi_1$$

$$\therefore\quad \psi_1 \text{ is an eigenfunction of } P_x^2 \text{ with an eigenvalue.}$$

$H^2 \cdot \frac{\pi^2}{L_x^2}$ or $\frac{h^2}{4L_x^2}$. Therefore, a series of measurements on identical system will

always give a precise result (eigenvalue).

$$P_x^2\ (state1) = \frac{h^2}{4L_x^2} = 2mE_1$$

$$\therefore P_x\ (state1) = \pm\sqrt{2mE_1}$$

But $\langle P_x \rangle = 0$

This interesting dilemma is resolved if we remember that Px represents the average of a large number of measurements of Px on identical sytem. A single measurement will

give$2 m\sqrt{E_1}$ or $-\sqrt{2mE_1}$ and in a series of measurements, $\sqrt{2mE_1}$ will be obtained as m- any times as is obtained. $-\sqrt{2mE_1}$ so that average is zero. We do not know in advance whether an expt. would give $+\sqrt{2mE_1}$ or $-\sqrt{2mE_1}$

∴ uncertainty in $P_x = P_x = 2\sqrt{2mE_1}$

$$= 2. H. \frac{\pi}{L_x} = \frac{h}{L_x}$$

$$= L_x$$

Maximum incertainty in position, x x

∴   . $P_x = L_x . \frac{h}{L_x} = h$ (*for state* $n = 1$)

.P

In general x x $= nh$

∴$_x . P_x \geq h$

This is Heisenberg's undertainty principle.

(b)   Consider the position x of the particle in the lowest state. The operator is

$x\$$

(multiplication by x)

$$\therefore x = \langle\rangle \quad \frac{2}{L_x} \int_{x\,0}^{L_x} x \sin^2 \frac{\pi x}{L_x} dx$$

$$\sin^2 a = \frac{1}{2}(1 - \cos 2a)$$

$$= \frac{2}{L_x} . \frac{1}{2} \int_0^{L_x} x\, 1 - \cos \frac{2\pi x}{L_x} dx$$

Now $\int_0^{L_x} x \cdot dx = \dfrac{x^2}{2}\Big|_0^{L_x} = \dfrac{L_x^2}{2}$

$\int_0^{L_x} x \cos \dfrac{2\pi x}{L_x} \, dx = \int_0^{L_x} u \, dv \qquad if \ u = x \qquad v = \dfrac{L_x}{2\pi L_x} \sin \dfrac{2\pi x}{}$

$\int_0^{L_x} = uv - \int v \, du$

$= x \cdot \dfrac{L_x}{2\pi} \sin \dfrac{2\pi x}{L} \Big|_{} - \int_0^{L_x} \dfrac{L_x}{2\pi} \sin \dfrac{2\pi x}{L} \, dx$

$= x \cdot \dfrac{L_x}{2} \sin \dfrac{2\pi x}{L_x} + \dfrac{L_x^2}{4\pi^2} \cos \dfrac{2\pi x}{L_x} \Big|_0^{L_x}$

$= 0 + \dfrac{L_x^2}{4\pi^2} - 0 + \dfrac{L_x^2}{4\pi^2} = \dfrac{L_x^2}{4\pi^2} - \dfrac{L_x^2}{4\pi^2} = 0$

It can be shown that $\langle x \rangle$ is independent of n, but

$\therefore \ \langle x \rangle = \dfrac{2}{L_x} \int_{0}^{L_x} x^2 \sin^2 \dfrac{n\pi x}{L} \, dx$

$= \dfrac{L_x^2}{3} \left[1 - \dfrac{3}{2n^2\pi^2} \right]$

(c)   Probability of finding the particle between x and x + dx is $\psi^2\, dx$. Using this we can find the probability of finding the particle between x = 0 and x = 1/4 Lx in   the ground state.

$$\int_0^{L_{x/4}} \psi^2_1 \, dx \qquad = \frac{2}{L_x} \int_0^{L_{x/4}} \sin^2 \frac{\pi x}{L_x} \, dx$$

$$= \frac{1}{L_x} \int_0^{L_{x/4}} 1 - \cos \frac{2x}{L_x} \, dx$$

$$= \frac{1}{L_x} \int_0^{L_{x/4}} dx - \int_0^{L_{x/4}} \cos \frac{2\pi x}{L_x} \, dx$$

$$= \frac{1}{L_x} \left[x - \frac{Lx}{2\pi} \sin \frac{2\pi x}{L_x} \right]_0^{L_{x/4}}$$

$$= \frac{1}{L_x} \left[\frac{L_x}{4} - \frac{L_x}{2\pi} - (0 - 0) \right]$$

$$= \frac{1}{4} - \frac{1}{2\pi} = 0.25 - 0.16 = 0.09 \qquad \text{Q } \frac{1}{\pi} = 0.32$$

(VI) Particle-in-a-box serves as a convenient model for several systems of interest e.g. $\bar{e}$ in a metal wire, π - electrons in linear conjugated molecules etc. It is possible to predict electronic spectra of conjugated polyenes.

The π-electrons are delocalized over the entire chain and may be regarded as particles in a one-dimensional box, chain of conjugated C atoms of length Nd where N is no. of C atoms and $d = \dfrac{1}{2}(d_{c-c} + d_{c=c})$

$$En = \frac{n^2 h^2}{8} mN^2 d^2 \; ; \psi = A \sin \frac{n\pi x}{Nd}$$

It is to be remembered that $N\pi$ - electros in the lowest state, fill the lowest N/2 levels (two in each level). Thus, absorption of energy would promote $\bar{e}$ from level with

$$n = \frac{N}{2} \quad to \quad n^1 = \frac{N}{2} + 1$$

$$\therefore E = hv = E_{\frac{N}{2}+1} - E_{\frac{N}{2}}$$

$$= (N+1)\frac{h^2}{8mN^2 d^2}$$

$$hv \approx \frac{h^2}{8mN^2 d^2}$$

$$\therefore v \, \alpha \, \frac{1}{N} \quad \text{i.e. frequency of absorption should be inversely propotional to}$$

chain length. This is borne out by the following data for polymers.

Compund	N	$\bar{v}(cm^{-1})$	$\lambda \overset{o}{A}$
Ethylme	2	61500	1625
Butadiene	4	46080	2170
Hexatriene	6	39750	2510

consequence of the symmetry properties of the system.

Chapter -4

SYSTEM WHERE THE POTENTIAL ENERGY IS NOT CONSTANT :

HYDROGEN ATOM :

(i) **One electron problem :** Hydrogen atom and hydrogen-like ions (He+, Li2+ etc) are the siplest of the chemical system with just a single electron outside the nucleus of charge + Ze. These are the only species for which the Schrodinger equation can be solved for EXACT SOLUTIONS. These solutions are called HYDROGENIC or ONE-ELECTRON WAVE FUNCTIONS (aromic orbitals); these from the basis fpr obtaining approximate solutions for more complex systems.

(ii) Special case of a two-body problem : The two particles the molecules and the electron - move about eacth other and interact electrostatically, the inter-action epending on the distance of separation, r

$$V(r) = - \frac{Ze_2}{r}$$

The mition of the relatively much heavier nucleus can be ignored (Born-Oppenheimer approximation) and problem reduced to motion of electron with reduced mass $\propto$ relative to the centre of mass of the system,

$$\propto = \frac{m_e . m}{N + \frac{N^m}{m_e}}$$

By virtue of very   (apherioally symmetric) large difference of messes
between proton and   electron

$\left(m_P \; 1846. \; m_e \right)$ the centre of mass practically coincides with the nucleus and thereforewe are justified in regarding the electron in these systems moving about a stationary nucleus fixed at the origin. Again

$$Q \quad m_P \;\; >> m_e \quad , \quad \propto \; \approx \; m_e$$

Since the potential has a spherical symetry, the problem may be regarded as a CENTRAL FIELD PROBLEM

Ignoring the nuclear motion (translational part) of the problem, the Schrodinger equation describing the internal motion of the hydrogen electron relative to the stationary proton can be writting

$$\vdots \quad -\frac{H^2}{2\alpha}\nabla^2 + V(r)\,\psi \quad = E\psi$$

Where $\quad V(r) = -\dfrac{e^2}{r}$ (hydrogen)

$$= -\frac{Ze^2}{r}\ \text{(hydrogen-like-ion)}$$

Solution of (2) will provide information about the possible energy states and spatial

distribution of the hydrogen electron. Eq. (2) is a three-variable equation which has been

$\dfrac{e^2}{r}$ privents the straight separation of x, y,

$\psi(xyz) = \psi_x(x).\psi_y(y).\psi_z(z)$ (r involves x, y, z, being related as r^2 $= x^2 + y^2 + z^2$)

However, remembering that $V(r)$ represents a centrosymmetric (spherical) potential

field, it seems logical to transform eq. (2) into spherical polar coordinates (r, θ, φ) where r becomes a single coordinate and needs no separation.

solved by separating the variables. $V(r) = -$ and z by writting.

r : distance of the point from the origin or the length of the line joining the point with the origin (OP)

$$0 \le r \le \infty$$

θ : angle between the line (OP) joining the pont to the origin and +Z axix

$$\le \theta \le \pi$$

φ : angle between the projection of the line OP in the XY plane and +X axis.

$$0 \le \varphi \le 2\pi$$

The relationship between cartesian coordinates and polar coordinates are :

$x = r \sin \theta \cos \varphi$ $\qquad$ $r = \left(x^2 + y^2 + z^2 \right)^{1/2}$

$y = r \sin \theta \sin \varphi$ $\qquad$ $\tan \theta = (x^2 + y^2)^{\frac{1}{2}} \, z$

$z = r \cos \theta$ $\qquad$ $\tan \theta = \; y$
x

$$d \, \tau = dx \, . \, dy \, .dz \quad \text{(cartesian coordinates)}$$

$$= r^2 \sin\theta \quad or \quad d\theta \; d\varphi \; \text{(polar coordination)}$$

$$\nabla^2 = \frac{\partial^2}{\partial x^2} \frac{\partial^2}{\partial y^2} + \frac{\partial^2}{\partial z^2} \qquad \text{(Cartesian coordination)}$$

$$= \frac{1}{r^2} \frac{\partial}{\partial x} r^2 \frac{\partial}{\partial t} + \frac{1}{r^2 \sin \theta} \frac{\partial}{\partial \theta} \sin\theta \frac{\partial}{\partial \theta} \cdot \frac{1}{r^2 \sin^2 \theta} \frac{\partial^2}{\partial \theta^2} \cdot$$

Equation (2) upon transformation into polar coordinates become

$$\frac{1}{r^2} \frac{\partial}{\partial r} r^2 \frac{\partial}{\partial r} + \frac{1}{r^2 \sin \theta} \frac{\partial}{\partial \theta} \sin 6 \frac{\partial \psi}{\partial 6} + \frac{1}{r^2 \sin^2 \theta}$$

$$+ \frac{8\pi^2 \propto}{1_2} \left[E - V (r) \right] \psi = 0$$

$$\psi = \psi(r\,\theta\,\varphi)$$

Separation of Variables :

Assume : $\psi = \psi(r\,\theta\,\varphi) = R(r).\theta(\theta).\Phi(\varphi)$

Substitute (4) in (3)

$$\frac{1}{r^2}\frac{\partial}{\partial r}\left(r^2\frac{\partial}{\partial r}\right)R\,\theta\,\Phi + \frac{1}{r^2\sin\theta}\frac{\partial}{\partial\theta}\left(\sin\theta\frac{\partial}{\partial\theta}\right)R\,\theta$$

$$+ \frac{1}{r^2\sin^2\theta}\frac{\partial^2}{\partial^2 6}R\,\theta\,\Phi + \frac{8\pi^2\alpha}{h^2}\left(E-V\right)R\theta\Phi = 0$$

$$\therefore \frac{\theta\Phi}{r^2}\frac{\partial}{\partial r}.r^2\frac{\partial R}{\partial r} + \frac{R}{r^2\sin\theta}\frac{\Phi\partial\sin 6}{\partial\theta}.\frac{\partial\theta}{\partial\theta} + \frac{R\theta\theta^2\Phi}{r^2\sin^2\theta}\,\partial\Phi^2$$

$$+ \frac{8\pi^2 \alpha}{h^2} \left(E - V \right) R\theta\Phi = 0$$

Devide by $R\,\theta\,\Phi$

$$\therefore \quad \frac{1\partial}{Rr^2\,\partial r} \cdot r^2 \cdot \frac{\partial R}{\partial r} + \frac{1}{\theta r^2 \sin\theta} \cdot \frac{\partial}{\partial\theta} \sin\theta \frac{\partial\theta}{\partial\theta} + \frac{1}{\Phi r^2 \sin^2\theta} \frac{\partial^2\Phi}{\partial\Phi^2}$$

$$+ \frac{8\pi^2 \alpha}{h^2} \left(E - V \right) = 0$$

multiply by $r^2 \sin^2\theta$

$$\therefore \quad \frac{\sin^2\theta}{R\partial r} \frac{\partial}{\partial r} r^2 \frac{\partial R}{\partial r} + \frac{\sin\theta}{\theta\partial\theta} \frac{\partial\theta}{\partial\Phi} + \frac{1}{\partial\Phi^2} \frac{\partial^2\Phi}{\partial\varphi^2}$$

$$+ \frac{8\pi^2 \alpha r^2 \sin^2\theta}{} \left(E - V \right) = 0$$

$$\therefore \quad \frac{1}{\Phi} \cdot \frac{\partial^2\Phi}{\partial\varphi^2} = -m^2 \qquad or, \qquad \frac{\partial^2\Phi}{\partial\varphi^2} + m^2\Phi = 0$$

$$\frac{\sin^2\theta}{R} \frac{\partial}{\partial r} r^2 \frac{\partial R}{\partial r} \quad \frac{\sin\theta}{\theta} \frac{\partial}{\partial\theta} \sin\theta \frac{\partial\theta}{\partial\theta} - m^2 + \frac{8\pi^2 \alpha r^2 \sin^2\theta}{h^2} \left(E - V \right) = 0$$

and

Devide (6) by $\sin^2\theta$ and rearrange

$$\frac{1}{R} \frac{\partial}{\partial r} r^2 \frac{\partial R}{\partial r} + \frac{8\pi^2 \alpha r^2}{h^2} \left(E - V \right) + \frac{1}{\theta\sin\theta} \frac{\partial}{\partial\theta} \sin\theta \frac{\partial\theta}{\partial\theta} - \frac{m^2}{\sin^2\theta} = 0$$

$$\qquad\qquad\qquad \text{depend on } r \qquad\qquad\qquad\qquad \text{depend on } \theta$$

$$-\frac{1}{R} \frac{\partial}{\partial r} r^2 \frac{\partial R}{\partial r} + \frac{8\pi^2 \alpha r^2}{h^2} \left(E - V(r) \right) = \beta$$

$$or \quad \frac{\partial}{\partial r} r^2 \frac{\partial R}{\partial r} - \beta R + \frac{8\pi^2 \alpha r^2}{h^2} \left(E - V \right) R = 0$$

$$\frac{1}{\theta \sin\theta} \frac{\partial}{\partial\theta} \sin\theta \frac{\partial\theta}{\partial\theta} - \frac{m^2}{\sin^2\theta} = -\beta \qquad or,$$

$$1 \quad \frac{1}{\theta \sin\theta} \frac{\partial}{\partial\theta} \sin\theta \frac{\partial\theta}{\partial\theta} - \frac{m^2}{\sin_2\theta}\theta + \beta\theta = 0$$

Equations (5), (7) and (8) each depends on only one variable. Thus we have

separated the variables and resolved the total three-variable eqn. (3) into three independent

Note : (i)   Energy of the system depends only on r but complete wave functions

will require solving all three equations.

(ii)   Although different two-particle systems will differ in their energy state and radial properties, they will have similar angular properties of their wave functions.

(iii)   We shall consider the three eqns. in the order

1. Φ eqn. 2. θ eqn. 3. R ean. To be more explicit we shall first find the Φ eqn.has acceptable solutions only for certain values of m. We shall introduce these values of m into θ eqn. and find that θ eqn. has acceptable solutions only for certain value of

β.Introducing these values of β into R eqn. we shall find that the acceptable solutions of R eqn. correspond to only certain value of E which constitute the allowed energy states of the system.

(1) Φ EQUATION

$$\frac{\partial_2\Phi}{\partial\varphi}2 + m^2\Phi = 0$$

The form of this eqn. is quite familiar to us. In fact, we have already discussed this eqn. and its solutions in some detail.

A satisfactory soln. of the eqn. is

$$\Phi_m(\varphi) = A\, e^{im\varphi}$$

In order that the solution is acceptable, it must be well-behaved. One of the well0behaved requirements is that the function must remain invariant on identity rotation. (i.e. complete rotation through 2π)

$$\therefore \Phi_m(\varphi) = \Phi_m(\varphi + 2\pi)$$

$$A\, e^{im\varphi} = A\, e^{im(\varphi + 2\pi)}$$

$$= A$$

$$e^{im\varphi} \cdot e^{2\pi im} \therefore e^{2\pi}$$

$$im = e^{-2\pi im} = 1$$

$$\therefore e^{2\pi im} + e^{-2\pi im} = 2$$

$$2\cos 2\pi m = 2 \quad or \quad \cos 2\pi m = 1$$

$$\therefore m = 0, \pm 1, \pm 2, \pm 3$$

Thus, for each numberical value of m (other than m = 0) there are two solutions of the form (9), one with positive m and the other with nehative m.

Thus we find that solution of Φ equation results in a restricted set of wave functions denoted by quantum nunber m which from its allowed values may be indentified with the magnetic quantum number of the Bohr ommerfeild theory.

The constant A in (9) may be obtained by using the normalization condition

$$\int_0^{2\pi} \Phi_m^* \, \Phi_m \, d\varphi = 1$$

$$\therefore A^* A \int_0^{2\pi} e^{im\varphi} e^{-im\varphi} \, d\varphi = A^* A \int_0^{2\pi} d\varphi$$

$$= A^* A\, 2\pi = 1$$

$$\therefore A^* A = |A|^2 = \frac{1}{2\pi} \qquad \therefore A = \pm \sqrt{\frac{1}{2\pi}}$$

Choosing the positive number the set of well-behaved solutios of Φ eqn. will be

$$\Phi_m(\varphi) = \frac{1}{2\pi} e^{im\varphi} \qquad m = 0, \pm 1, \pm 2, \ldots\ldots\ldots\ldots\ldots$$

The solutions we have discussed so far have been complex solution. For some purpose, it is convenient to have alternative real solutions. These can be readily obtained

since if $\Phi_{|m|} = A\, e^{im\varphi}$ $\Phi_{-m} = A\, e^{-im\varphi}$ and are solutions of the eqn. then linear combinations of these functions will also be the solutions. Thus for a given $|m|$, the sum and the difference of two complex solutions will give

$$\Phi_{|m|}(\varphi) = \frac{1}{\sqrt{\pi}} \cos m\,\varphi$$

$$\Phi_{|m|}(\varphi) = \frac{1}{\sqrt{\pi}} \sin m\,\varphi$$

For m=0, there is only one real solution.

The Φ -solutions, except for m=0, are all complex. For obtaining them in real form, the two solutions corresponding to the same $|m|$ are linearly combined i.e. added and subtacted. For example.

$$\Phi_1 \qquad + \qquad N\left(\Phi_1 + \Phi_{-1}\right) = \frac{1}{\sqrt{\pi}} \cos\varphi$$

$$\Phi_{-1} \qquad - \qquad N\left(\Phi_1 - \Phi_{-1}\right) = \frac{1}{\sqrt{\pi}} \sin\varphi$$

Consider $N\left(\Phi_1 + \Phi_{-1}\right)$:

$$\int_0^{2\pi} N\,(\Phi_1 + \Phi \qquad {}^* \quad N\,(\Phi_1 + \Phi \qquad d\varphi = 1$$

$$\therefore \quad N^{*}N \int_0^{2\pi} \Phi_1^{*}\Phi_1 \, d\varphi + \int_0^{2\pi} \Phi_1^{*}\Phi_{-1} \, d\varphi + \int_0^{2\pi} \Phi_{-1}^{*}\Phi_1 \, d\varphi + \int_0^{2\pi} \Phi_{-1}^{*}\Phi_{-1} \, d\varphi = 1$$

Since Φ_1 and Φ_{-1} are individually normalized and as a pair orthigonal

$$\therefore N^{*}N \left[1 + 1 \right] = 1$$

$$\therefore N^{*}N = |N|^2 = \frac{1}{2}$$

$$\therefore N = \frac{1}{\sqrt{2}}$$

$\therefore$ Normalized linear combination is

$$\frac{1}{\sqrt{2}} \left(\Phi_1 + \Phi_{-1} \right)$$

$$= \frac{1}{\sqrt{2}} \cdot \frac{1}{\sqrt{2\pi}} \left(e^{i\varphi} + e^{-i\varphi} \right)$$

$$= \frac{1}{\sqrt{2}} \cdot \frac{1}{\sqrt{\pi}} \cdot 2 \cos\varphi$$

$$\frac{1}{\sqrt{\pi}} \cdot \cos\varphi$$

Similarly, consider $N \left(\Phi_1 - \Phi_{-1} \right)$

$$\therefore N \cdot \frac{1}{\sqrt{2\pi}} \left(e^{i\varphi} - e^{-i\varphi} \right) = N \cdot \frac{1}{\sqrt{2\pi}} \cdot 2i \sin\varphi$$

$$= \frac{2i\,N}{\sqrt{\pi}} \sin\varphi .$$

Evaluate N by normalization

$$\therefore \quad \frac{2i^2 N^{*}N}{\sqrt{\pi}} \int_0^{2\pi} \sin^2\varphi \, d\varphi = 1$$

$$\therefore \quad \frac{2i^2 N^* N}{\pi} \cdot \frac{1}{2} \int_0^{2\pi} 2\pi \, d\varphi - \int_0^{2\pi} \cos 2\varphi \, d\varphi \quad = 1$$

$$\frac{2\pi}{2} - \left[\sin 2\varphi\right]_{0} = 0$$

$$\therefore \; 2\,i^2\,N^*\,N = 1 \qquad i.e.\; N^*\,N = |N|^2 = \frac{1}{2}\,i^2$$

$$N = \frac{1}{\sqrt{2}\,.i}$$

$$\therefore \quad N\left(\Phi_1 - \Phi_{-1}\right) = \frac{1}{\sqrt{\pi}} \sin \varphi$$

In general

$$
\Phi_1 \bigg\rangle \quad + \quad \bigg\langle \quad \frac{1}{\sqrt{\pi}} \cos|m|\varphi
$$

$$
\Phi_{-1} \bigg/ \quad - \quad \bigg\langle \quad \frac{1}{\sqrt{\pi}} \sin|m|\varphi
$$

SOME $\Phi_m(\varphi)$ FUNCTIONS

m	function	in terms of φ	
0	$\Phi_0(\varphi)$	$\dfrac{1}{\sqrt{2\pi}}$	
1	$\Phi_1(\varphi)$	$\dfrac{1}{\sqrt{2\pi}}e^{i\varphi}$	$\dfrac{1}{\sqrt{\pi}}$
-1	$\Phi_{-1}(\varphi)$	$\dfrac{1}{\sqrt{2\pi}}e^{-i\varphi}$	$\dfrac{1}{\sqrt{\pi}}$
+2	$\Phi_2(\varphi)$	$\dfrac{1}{\sqrt{2\pi}}e^{2i\varphi}$	$\dfrac{1}{\sqrt{\pi}}$
-2	$\Phi_{-2}(\varphi)$	$\dfrac{1}{\sqrt{2\pi}}e^{-2i\varphi}$	$\dfrac{1}{\sqrt{\pi}}$

(2) **θ EQUATION** :

$$
\frac{1}{\sin\theta}\frac{d}{d\theta}\sin\theta\,\frac{d\theta}{d\theta} - \frac{m^2}{\sin^2\theta}\theta + \beta\,\theta = 0
$$

The solutions to this eqn. are not obvious as in the case of Φ eqn. is of a form familiar to mathematicians as the Legendre's equation for which solutions were well-known long before the advent of Quantum mechanics. The solutions are associated Legendre functions generally denoted by $P_r^m(\cos\theta)$. These have specfic functional form for given values of l and m.

Eq. (8) may be written

$$\frac{1}{\sin\theta}\frac{d}{d\theta}\sin\theta\frac{d\theta}{d\theta} + \beta - \frac{m^2}{\sin^2\theta}\theta = 0$$

taking the indicated derivative

$$\frac{d^2\theta}{d\theta^2} + \frac{\cos\theta}{\sin\theta}\frac{d\theta}{d\theta} + \beta - \frac{m^2}{\sin^2\theta}\theta = 0$$

θ must be well-known in the interval $\qquad 0 \le \theta \le \pi$

or $\qquad -1 \le \cos\theta \le 1$

Let us change the variable as

$$x = \cos\theta$$

$$\therefore \quad \frac{d}{d\theta} = -\sin\frac{d}{dx}; \quad \frac{d^2}{d\theta^2} = \sin^2\theta\frac{d^2}{dx^2} - \cos\theta\frac{d}{dx}$$

Substitute into 915) and simplify

$$\therefore \quad \sin^2\theta\frac{d^2\theta}{dx^2} - 2\cos\theta\frac{d\theta}{dx} + \beta - \frac{m^2}{\sin^2\theta}\theta = 0 \qquad \text{or,}$$

$$\left(1-x^2\right)\frac{d^2\theta}{dx^2} - 2x\frac{d\theta}{dx} + \beta - \frac{m^2}{\sin^2\theta}\theta = 0$$

The well-known associated Legendre's equation is

$$\left(1-x^2\right)\frac{d^2\alpha}{dx^2} - 2x\frac{d\alpha}{dx} + 1\left(l+1 - \frac{m^2}{1-x^2}\right)\alpha = 0 \qquad l = 0, 1, 2, 3, \ldots\ldots\ldots$$

$$\alpha = P_l^m(x) = 1-x^{m}\left(\right)^2\frac{d^m}{dx^m}P_l(x)$$

Where $P_l(x)$ is the Legendre's polynomical of degree l.

$$P_l(x) = \frac{1}{2^l \, l^l} \frac{d^l}{dx^l} \left(x^2 - 1 \right)^l$$

$$= \frac{\left(\frac{-1}{2^l \, l^l} \right)^l \frac{d^l}{dx^l} \left(1 - x^2 \right)^l}{}$$

$P_l^m(x)$ are orthogonal in the interval $\quad -1 \le x \le 1$

and $P_l^m(x) = m$ for $m > 1$

Normalized assoc. legendre functions are

$$\alpha(x) = \frac{(-1)^l}{2^l \, l!} \sqrt{\frac{2l+1}{2} \cdot \frac{(1-|m|)!}{(1+|m|)!}} \, (1-x^2)^{|m|/2} \frac{d^{l+|m|}}{dx^{l+|m|}} (1-x^2)^l$$

Our θ eqn. (18) becomes identical with the associated Legendre's eqn. (19) if we put

$$= l(l+1)$$

and then the θ-solutions are given by (21). Since m enters θ-eqn. as m^2.

$$\theta_{lm}(\theta) = \theta_{l-m}(\theta)$$

Normalized θ solutions have the form

$$\theta_{lim}(\theta) = \frac{-1^l}{2^l \, l!} \sqrt{\frac{2l+1}{2} \cdot \frac{(1-|m|)!}{(1+|m|)!}} \sin^{|m|}\theta \, \frac{d^{l+|m|}}{dx^{l+|m|}} \sin^{2l}\theta$$

For $\quad \theta_{1\pm l}(\theta) \quad l=1 \quad |m|=1$

$$\theta_{1\pm l}(\theta) = \frac{(-1)^l}{2^l \, l!} \sqrt{\frac{3}{2} \cdot \frac{(1-1)^l}{(1+1)!}} \sin\theta \, \frac{d^2}{d\cos\theta^2} \sin^2\theta$$

$$\sin^2\theta = 1 - \cos^2\theta$$

$$\therefore \quad \frac{d}{d(\cos\theta)}(1-\cos^2\theta) = -2\cos\theta; \quad \frac{d^2}{d(\cos\theta)^2}(1-\cos^2\theta) = -2$$

$$\theta_{1\pm l}(\theta) = -\frac{1}{2}\sqrt{\frac{3}{4}}\sin\theta(-2) = \sqrt{\frac{3}{4}}\sin\theta$$

SOME $\theta_{l \pm m}(\theta)$ FUNCTIONS

$\underline{l}$	$\underline{m}$	Function
0	0	$\theta_{00}(\theta) = \dfrac{\sqrt{2}}{2}$ or $\dfrac{1}{\sqrt{2}}$
1	0	$\theta_{10}(\theta) = \dfrac{\sqrt{6}}{2} \cos\theta$
	± 1	$\theta_{1 \pm 1}(\theta) = \dfrac{\sqrt{3}}{2} \sin\theta$

2	0	$\theta_{20}(\theta) = \dfrac{\sqrt{10}}{4}\left(3\cos^2\theta - 1\right)$
	±1	$\theta_{2\pm1}(\theta) = \dfrac{\sqrt{15}}{2}\sin\theta\cos\theta$
	±2	$\theta_{2\pm2}(\theta) = \dfrac{\sqrt{15}}{4}\sin^2\theta$

Note that solution of θ eqn. introduces a new quantum number l (l = 0, 1, 2, 3, ...) which may be indentified.

with the new azimuthal or orbital momentum quantum number of Boht-Sommerfeld theory, Also it restricts m to

$$m = 0, \pm 1, \pm 2, \ldots\ldots\ldots\ldots, \pm 1$$

(3)   R-equation :

$$\frac{1}{r^2}\frac{d}{dr}r^2\frac{dR}{dr} - \frac{\beta}{r^2}R + \frac{2\alpha}{H}\left(E - V(r)\right)R$$

$$V(r) = -e^2/r \qquad \text{(hydrogen)}$$

$$= -Z\,e^2/r \qquad \text{(hydrogen-like ion)}$$

and $\beta = 1(1 + 1)$; $1 = 0, 1, 2,\ldots\ldots$

$$\therefore \quad \frac{1}{r^2}\frac{d}{dr}r^2\frac{dR}{dr} + -\frac{l(l+1)}{r^2} + \frac{2\alpha}{H}\frac{Ze^2}{r}R = 0$$

Introduce $a^2 = -2\alpha E/H^2$ and $\lambda = \alpha \dfrac{Ze^2}{H^2}a$

and change the variable $r \to p : p = 2\,\alpha\,r$

$\therefore$ Eq. (27) becomes

$$\frac{d^2R}{dp^2} + \frac{2}{p}\frac{dR}{dp} + \frac{-l(l+1)}{p^2} - \frac{1}{4} + \frac{\lambda}{p}\,R = 0$$

$$R = R(p)$$

Eq. (29) has been solved by a POLYMINAL or POWER SERIES METHOD and the solutions are the well-known associated langerre polynomicals :

$$R_{nl}(p) = N_{nl} \cdot G(p) \cdot e^{-\frac{1}{2}p}$$

$G(p)$ are related to (for given n and l) assoc. Langurre polyminals as

$$G(p) = pl \cdot L_{n+l}^{2l+1}(p)$$

$L^s{}_r(p)$ is defiend as

$$L^s{}_r(p) = \frac{d^s}{dp^s} L_r(p)$$

Where $L_r(p)$ is the Langurre polynomial of degree r given by

$$L_r(p) = e^p \frac{d^r}{dp^r}\left(p^r e^{-p}\right)$$

$$\therefore R_{nl}(p) = N_{nl} \cdot pl \, L_{n+l}^{2l+1}(p) \; e^{-\frac{1}{2}p}$$

The normalization factor N_{nl} is given by

$$N_{nl} = -\frac{2Z^3}{na_0} \frac{(n-l-1^i}{2n \, (n+1^i)^3}^{\frac{1}{2}}$$

Note :

(i) while obtained well-known solutins $R(p)$, λ turns out to be a non-zero positive integer and this may be identified with the principal quantum number, n.

$$\lambda = n = 1, 2, 3, \ldots\ldots\ldots\ldots, \infty \qquad (n \neq 0)$$

Also, the values of l are restricted as

$$l = 1, 2, \ldots\ldots\ldots\ldots (n - 0)$$

(ii)   The familiar Bohr expression for quantized energy states of a hydrogen-like system follows at once from the definition of $\lambda = n$ on substitution for α.

Let us construct $R_{10}(r)$:

$$n = 1 \qquad l = 0 \qquad\qquad \therefore$$

$L^1_1(p)$ is required

$$L_1(p) = e^P \ d\left(p\, e^{-P}\right) /$$

$$dp = 1 - p$$

$$\therefore L_1(p) = d\, L_1^{\ (p)}\, dp = d^{\left(1_- p\right)}\, dp =$$

$$-1$$

$$\therefore G\ (/p) = p^0\ (-1) =$$

$$-1$$

$$\text{Now} \quad {}^{IV}_{10} = -\frac{2Z}{a\,0}^{3} \frac{oi}{2\,1i}^{3}{}^{\frac{1}{2}} = -\frac{2Z}{a0}^{3}\, {}^{1}\cdot\frac{1}{2}^{\frac{1}{\angle}-}$$

$$()$$

We have seen that complete solution of the wave equation for well-behaved solns. introduces three quantum nos. n, l, and m. These quantum nos. were arbitrarily introduced in the Bohr Somerfeld theory whereas in wave mechanics they appear as a consequence of a few basic postulates of wave mechanics.

The total wave function descriing any state of H atom is, of course, the product of R, θ and Φ functions

$$\psi_{nlm}(r\theta\varphi) = R_{nl}(r)\,\Theta_{lm}\,\Phi_m(\varphi)$$

$$= R_{nl}(r)\,Y_{lm}(\theta\varphi)$$

ψ_{nlm} sis referred to as the SPACE WAVE FUNCTION (ORBITAL)

The complete hydrogen atom wave functions are listed below :

H-LIKE WAVE FUNCTION

$$\sigma \quad \frac{n}{2}p = \frac{z}{a_0}r$$

K-SHELL

$$n = 1 \qquad l = 0 \qquad m = 0 \qquad \psi_{1s} = \frac{1}{\sqrt{\pi}}\left(\frac{Z}{a_0}\right)^{\frac{3}{2}}e^{-\sigma}$$

K-SHELL

$$n = 2 \qquad l = 0 \qquad m = 0 \qquad \psi_{2s} = \frac{1}{4}(2\pi)^{-\frac{1}{2}}\left(\frac{Z}{a_0}\right)^{\frac{3}{2}}(2-\sigma)e^{-\frac{\sigma}{2}}$$

$$n = 2 \qquad l = 1 \qquad m = 0 \qquad \psi_{2p2} = \frac{1}{4}(2\pi)^{-\frac{1}{2}}Z^{\frac{3}{2}}\sigma e^{-\frac{\sigma}{2}}\cos\theta$$

$$n = 2 \qquad l = 1 \qquad m = \pm 1 \qquad \psi_{2px} = \frac{1}{4}(2\pi)^{-\frac{1}{2}} Z {\left(\frac{u_0}{u_0}\right)}^{\frac{3}{2}} \sigma e^{-\frac{\sigma}{2}} \sin\theta \cos\varphi$$

$$\psi_{2py} = \frac{1}{4}(2\pi)^{-\frac{1}{2}} \left. Z \middle/ a_0 \right.^{3/2} \sigma e^{-\sigma/2} \sin\theta\cos\theta$$

M-SHELL

$n = 3 \qquad l = 2 \qquad m = 0 \quad \psi_{3s} = \frac{1}{81}\sqrt{3\pi} \left. \middle/ a_0 \right.^{3/2} (27 - 18\sigma + 2\sigma^2) e^{-\sigma/3}$

$n = 3 \qquad l = 1 \qquad m = 0 \quad \psi_{3pz} = \frac{\sqrt{2}}{81}\sqrt{\pi} \left. Z \middle/ a_0 \right.^{3/2} \sigma(\sigma - \sigma) e^{-\sigma/3} \cos\theta$

$n = 3 \qquad l = 1 \qquad m = \pm \quad \psi_{3px} = \frac{\sqrt{2}}{81}\sqrt{\pi} \left. Z \middle/ a_0 \right.^{3/2} \sigma(\sigma - \sigma) e^{-\sigma/3} \sin\theta\cos\varphi$

$$\psi_{3py} = \frac{\sqrt{2}}{81}\sqrt{\pi} \left. Z \middle/ a_0 \right.^{3/2} \sigma(\sigma - \sigma) e^{-\sigma/3} \sin\theta\cos\theta$$

$n = 3 \qquad l = 2 \qquad m = 0 \quad \psi_{3dz^2} = \frac{1}{81}\sqrt{6\pi} \left(Z/a \right)^{3/2} \sigma^2 e^{-\sigma/3} (3\cos^2\theta - 1)$

$n = 3 \qquad l = 1 \qquad m = \pm \quad \psi_{3dxz} = \frac{\sqrt{2}}{81}\sqrt{\pi} \left. Z \middle/ a_0 \right.^{3/2} \sigma^2 e^{-\sigma/3} \sin\theta\cos\theta\cos\varphi$

$$\psi_{3dyz} = \frac{\sqrt{2}}{81}\sqrt{\pi} \left. Z \middle/ a_0 \right.^{3/2} \sigma^2 e^{-\sigma/3} \sin\theta\cos\theta\sin\varphi$$

$n = 3 \qquad l = 2 \qquad m = \pm 2 \quad \psi_{3dx^2} = \frac{1}{81}\sqrt{2\pi} \left. Z \middle/ a_0 \right.^{3/2} \sigma^2 e^{-\sigma/3} \sin^2\theta\cos^2\varphi$

$$\psi_{3dxy} = \frac{1}{81}\sqrt{2\pi} \left. Z \middle/ a_0 \right.^{3/2} \sigma^2 e^{-\sigma/3} \sin^2\theta\sin^2\varphi$$

SCF METHOD

SELF CONSISTENT FIELD METHOD :

DRAWBACK OF HUCKEL THEORY (LIMITATION)

The main limitation of Huckel theory was that, theory did not consideration of the $\bar{e}$ repulsion energy term $\frac{e^2}{r_{ij}}$ like which depends on the co-ordinate of both $\bar{e}$. Due to this, the thory failed to explain the excited state phenomena. However in a closely related series of molecules this neglet of $\bar{e}$ repulsion will be minimum and so gives statis factory results, (Hukels theory $\bar{e}$ repulsion)

In the self consistent field method a single $\bar{e}$ is treated as if it was moving in the presence of a central electrical field resulting from the average charge distribution of the nucleus and the remaining $\bar{e}$ s. This is to estimate the potential energy function due to the nucleus and all of the $\bar{e}$ s.

The solution to the wave eqn. for the first $\bar{e}$ will give the average central field which can be used for the wave equation of the second $\bar{e}$ and so on this procedure gives improved wave function for the $\bar{e}$ untill no appreciable change is noted. At which point the field is set to be self-consistent.

For a $\wedge$ -$\bar{e}$ system, the humiltoniam operator is expressed in atomic units as

$$\bar{H} = \sum^{all\ \bar{e}} \bar{H}_i (core) + \frac{1}{2} \sum_{i \neq j} \frac{1}{r_{ij}} \quad\text{.................(1)}$$

i & j represents $\bar{e}$ pair $\frac{1}{r_{ij}} = \bar{e}$ repulsion term.

The first term $\bar{H}_i$ (core) determines the attraction energy of the core for teh $\wedge$ - electron, which is given by

$$\bar{H}_i (core) = -\frac{1}{2} \nabla_i^2 - \sum^{all\ nuclei\ a} \frac{Z_a}{r_{ia}} \quad\text{..............(2)}$$

Where Za the nuclear charge on nucleus ria = the distance of an $\bar{e}$ " i "^{a} from nucleas "a"

To evaluate $\frac{1}{r_{ij}}$ lets first take the case of $2\bar{e}$ s. $\bar{e}$ 1st is subjected to the average field of an $\bar{e}$

(2) which occupies the molecular orbital Ψi

The average field is given by

$$\int \Psi_{j^2 (2)} \left(\frac{1}{r_{12}} \right) dT_2 \quad \text{.............(3)}$$

Since the $\bar{e}$ can exchange bet^n the

M.O. " Ψi" & " Ψj" this can be represented as

$$\int \Psi_{j^2 (2)} \Psi j_{(2)} \left(\frac{1}{r_{12}} \right) dT_2 \quad \text{.............(4)}$$

The eigen value eq^n to be solved for an $\bar{e}$ (1) in molecular orbital say " Ψj" is given by

$$\left[\sum_{i}^{all} \hat{H}_i (core) + \sum^{occ.Mos} (2\hat{J}_i - \bar{K}_i) \right] \Psi j_{(1)} = \Sigma j \Psi j_{(1)} \quad \text{.......(5)}$$

Where, $\quad \hat{J} i \Psi i_{(1)} = \{ \int \Psi i^2(1) \left(\frac{1}{r_{12}} \right) dT_2 \} \Psi j_{(1)}$

$$\bar{K} i \Psi i_{(1)} = \{ \int \Psi i(2) \Psi j (2) \left(\frac{1}{r_{12}} \right) dT_2 \} \Psi j_{(1)}$$

$\hat{J}$ = coulomb operator

$\bar{K}$ = Exchange operator

The L.H.S. of the eq-(2) is called as the effective Fock- operator ($\hat{F}$) and the eigen value eq^n is simply written as

$F \Psi_{J(1)} = \epsilon j \Psi_{J(1)}$

$\hat{F} \Psi i = \epsilon i \Psi_i$ ....(6)

As in simple Huckel theory, the M.O. built up in the LCAO form as

$\Psi i = \sum_{j}^{all\ MOs} C_{ij} \phi_j$ ....(7)

Application of variation method leads to a set of secular equations the determinant of which can be written as

$$
\begin{vmatrix}
\alpha_{11}\text{-}S_{11}E & \beta_{12}\text{-}S_{12}E & \beta_{13}\text{-}S_{13}E \\
\beta_{21}\text{-}S_{21}E & \alpha_{22}\text{-}S_{22}E & \beta_{23}\text{-}S_{23}E \\
\text{------} & \text{-------} & \text{--------} \\
\text{-----} & \text{-------} & \text{--------}
\end{vmatrix} = 0
$$

$$\alpha i = F\, ii = <\phi_i|F|\phi_i>$$

$$\beta ij = Fij = <\phi_i|F|\phi_j> \qquad \dots\dots\dots\dots(9)$$

$$sij = <\phi i\,/\,\phi j>$$

In integral αi & βij can be evaluated using the relationship

$$\alpha i = \alpha i\,(\text{core}) + \frac{1}{2} <\phi i\,\phi i\,/\,\phi i\,\phi i> + \sum_{i\ne j} q_i <\phi i\,\phi i\,/\,\phi j\,\phi j> \quad \dots\dots\dots\dots(10)$$

$$\beta_{ij} = \beta_{ij}\,(\text{core}) - \frac{1}{2} P_{ij} <\phi i\,\phi i\,/\,\phi j\,\phi j> \quad \dots\dots\dots(11)$$

Here, q_i = total $\overline{\Lambda}\,\bar{e}$ s density on atom " i "

P_{ij} = Bond order

$\alpha_i(\text{core})$ and $\beta_{ij}\,(\text{core})$ are given by the Expression.

$$\alpha_i\,(\text{core}) = <\phi_j\,/\,\sum_i^{all} H_i\,(\text{core})\,/\,\phi_j>, \, i = j$$

$$\beta_{ij}\,(\text{core}) = <\phi_i\,/\,\sum_i^{all} H_i\,(\text{core})\,/\,\phi_j>, \, i \ne j$$

$$\ne 0. \text{ atom i being linked to atom j}$$

$$= 0, \text{ otherwise}$$

With the above Eq^n (10) $< \phi i \, \phi i \, / \, \phi i \, \phi i >$ describes the electron repulsion between two $\bar{e}$ s in the field of the same core and $< \phi i \, \phi i \, / \, \phi j \, \phi j >$ in Eq^n (11) between two $\bar{e}$ s on the neighbouring cores ϕi and ϕj with this eq^{ns} the cofficients of the secular diterminant is solved & finally the $\bar{e}^{\,s}$ density **q** & the bond order **P** are calculated.

These set of q & p values are used to calculated an improved set of α & β values and the process is continued until consistency is achieved.

The self consistency formula of M.O. theory is based on may $\bar{e}$ Hamiltonian developed " Roothan" incorporating the $\bar{e}$ repulsion energy term $\frac{1}{r_{ij}}$ which depends on the co-ordinate of both $\bar{e}$ s which cannot be repleaced by a one $\bar{e}$ potential energy term. To calculate α_i & β_{ij} we should have the values of α_i (core) i.e. the one centre $\bar{e}$ repulsion term and β_{ij} i.e. the two centred $\bar{e}$ repulsion term to evaluate these two we have to adopt the ppp approximation

(PPP Apro^n α_i (core), β_{ij} (core))

Chapter- 5

APPROXIMATION MATHOD

Wave equation can not be solved directly but by Approximation method. There are two approximation method.

    i)   Variation method

    ii)   Parurbation method

APPROXIMATIONA METHOD :

Variation and Perturbation

I   Variation Method :

Shrodinger's Wave equation

$$\nabla^2 \psi + \frac{8\pi^2 m}{H^2} (E - V) \psi = 0$$

ψ = wave function of single particle

$$\frac{-H^2}{8\pi^2 m} \nabla^2 + V\psi = E\psi$$

$$H\psi = E\psi$$

H = Hamiltanion operator

Operator instruction

$$E = \frac{H\,\psi}{\psi}$$

$$E = \frac{H\,\psi\,\psi^*}{\psi\,\psi^*}$$ to make constant

integration over all the co-ordinates give equation to calculate energy at any position of particle.

$$E = \frac{\int \psi H \psi^*}{\int \psi\psi^*}$$

ψ is normalised

$$E = \langle \psi H \psi^* \rangle$$

assume certain value for ψ to solve the wave equation.

$\psi = \psi_1$ trial wave function 20

$\psi = \psi_2$ trial wave function 5 true wave function

$\psi = \psi_3$ trial wave function 15 cause min energy

Statement :

The variation theory can be stated as follow

if ψ is the trial wave function of a system whose hamiltanion a discrete eigen spectrum

then $\langle \psi | H | \psi \rangle \geq E_0$.

Where E_0 is the lowest exact eigen value of H^+ . if trial wave function ψ is normalised.

Proof : Choose d trial wave function :

If φ_0 , φ_1 , φ_2 are sets of nimalised and mutually orthogonal eigen function of operator H^+ with discrete eigen value. E_0 , E_1 , E_2 we can write

$$H \varphi i = E \varphi i ,\qquad i = 0, 1, 2, 3$$

to prove the theoram,

We have to expand, ψ in terms of these equation function.

$$\psi = \sum_i ci\, \varphi i \qquad i = 0, 1, 2$$

Since ψ is normalised

$$\langle \psi / \psi_j \rangle = 1$$

$$\sum_i c_i^2 = 1$$

Equation for the calculation of energy :

$$\langle \varphi | H | \varphi_{ij} \rangle = E_i$$

and also

$$\langle \varphi_i | H | \varphi_j \rangle = E_j \langle \varphi_i / \varphi_j \rangle = 0$$

consider inc. integral.

$$E = \langle \varphi | H | \varphi \rangle$$

$$= \left\langle \sum_i c_i \varphi_i | H | \sum_i c_j \varphi_j \right\rangle$$

$$= \sum_i c_i^2 \langle \varphi_i | H | \varphi_i \rangle + \sum_{i \neq j} c_i c_j \langle \varphi_i | H | \varphi_j \rangle$$

Minimize the E are variable parameter

$$\langle \varphi | H | \varphi_j \rangle = E_i j \langle \varphi_i | H | \varphi_j \rangle = 0$$

$$E = \sum_i c_i^2 E_i$$

If E_0 is also energy. then we can write

$$E - E_0 = \sum_i c_i^2 E_i - E_0$$

$$= \sum_i c_i^2 E_i - \sum_i c_i^2 E_0$$

$$= \sum_i c_i^2 (E_i - E_0)$$

Since $\sum_i c_i^2$ is always positive

$E_i - E_0$ must be positive or zero for all value of i therefore H follows that

$$E_i - E_0 \geq 0$$

or $\quad \langle \varphi | H | \varphi \rangle \geq 0$

if the trial ware function φ happens to be true wave function of the system in its lowest energy state, we get the true energy.

The application of variation method involve following step.

(1)   Choose a try wave function with some variable parameter.

(2)   Calculation of the energy $\langle \varphi | H | \varphi_l \rangle$ integral.

(3)   Minimize the integral w.r.t. variable parameter.

(4)   The function φ with the optimum value is the best approcimation.

TO CALCULATE THE ENERGY :

$$E = \frac{\int \psi \, H \, \psi \cdot d}{\lambda \int \psi^2 \cdot d\lambda}$$

We have to get limit of integration the e^- can be consider to be in spherical shell at a distance r to $r + dr$ from the nucieous as H goes from 0 to 2 since volume of the spherical sheel is $4\pi r^2 \, dr$. $d\lambda$ = Volume element can be substituted with this value.

substituting this value in equation (4)

$$E = \int_0^2 e^{-ar} \left[\frac{H^2}{8\pi^2 \, m} \frac{a^2 - 2a}{r} e^{-ar} - \frac{e^2}{r} e^{-ar} \right] 4\pi r^2 \, dr$$

$$= \int_0^2 e^{-2ar} \, 4\pi r^2 \, dr$$

This equation (8) can be solve as

$$= \int_0^2 \frac{-h^2 a^2}{8\pi^2 \, m} r^2 e^{-2ar} \, dr + \int_0^2 \frac{2h^2 a^2}{8\pi^2 \, m} r^2 e^{-2ar} \, dr - \int_0^2 e^2 r e^{-2ar} \, dr$$

$$= \int_0^2 r^2 e^{-2ar} \, dr$$

This is a γ-function for which solution is known. solution for this integral is by adpting γ-function.

$$E = \dfrac{\dfrac{-h^2 a^2 2}{8\pi^2 m \, 8a\,2} + \dfrac{2h^2 a}{8\pi^2 m}\dfrac{1}{4a^2} - e^2\dfrac{1}{4a^2}}{\dfrac{2}{8a^2}}$$

$$= \dfrac{-}{} \dfrac{h_2 a_2}{8\pi^2 m} - e^2 a$$

MINIMISE E :

To minimize the energy w.r. to variable a partically diffrenciate F a.r to a

$$\therefore \dfrac{2E}{m} = \dfrac{2h\,a}{} {}^2 - e^2 a = 0 \; 2a\,8\pi^2$$

$$\therefore a = \dfrac{4\pi^2 m\,e^2}{h^2}$$

Substituting this value of (a) in equation (11). we get

$$E_{min} = \dfrac{-2\pi^2 m\,e^4}{h^2}$$

by directly solving equation

$$E = \dfrac{1}{2}a.a \text{ (closed from calculation)}$$

Which is the same as the energy calculated by directly solving the wave equaton. If we would have choosen

(trial wave function)

The value of energy $E = 0.424$ *a.u.*

$\psi = e^{-a^2 r}$ Energy E is slightly higher than the energy value E calculated for $\psi = e^{-ar}$ which means that $\psi = e^{-ar}$ is more closer to the exact value of E and so is the true wave function.

PERTURBATION METHOD :

Energy levels like H or He

Small disturbance

based on pentribation theoram

(1)   Time independent or stationaty

(a) non-degencrate (b) Degencrate - same energy

(2)   Time dependant

TIME INDEPENDANT OR STATINARY :

$H = H^{(0)} + H^{(1)}$ - Small and poruted porburation Hamiltanion. Large unperturbed hamiltanion

This theory help in poiting changes in energy levels and eigen function of Cl system caused by small disturbance in such case the hamilanion H can be consider to be made of two parts.

(1) Which is large i charactorises a system for which wave function equation can be solved exactly that is unperturbed $H^{(0)}$.

(2) Which is small and is relative to the perturbed state $H^{(1)}$ perturbation hamiltanian H can be represented as,

$$H = H^{(0)} + H^{(1)}$$

in the perturved state. the wave equation can be written as $H \psi = E \psi$ if $\psi n^{(0)}$ I

$En^{(0)}$ are the orthonormal eigen function I eigen values of unpenturbed hamitaniun $H^{(0)}$.

$H^{(0)} \psi_n^{(0)}$ $= En^{(0)} \psi_n^{(0)}$ unperturved state

PERTURBATION THEORY OF NON-DEGENRATE SYSTEM:

non-degenrate system means that there will be one ideal function corresponding to each energy level or eigen value.

if there are two or more eigen function correspond to one energy level, this system is called as a degenerate system.

To solve any problem by perturbation theory is should satisfied the following condition.

(1) The H can be devided into two parts

$$H = H^{(0)} + H^{(1)}$$

(2) The eigen value d eigen functin of unperturbed $H^{(0)}$ are known

$$H^{(0)} >> H^{(1)}$$

(3) The wave equation for perturbed system presented as,

$$H \psi = E \psi$$

Porturbed state

in this stationary state H does nor depends on time d it is possible to expand H in term of expansion parameter π as,

$$H = H^{(0)} + \lambda H^{(1)} + \lambda^2 H^{(2)}$$

Were $H^{(0)} >> H^{(1)}$

Suppose the eigen valued and equal function cd unperturable system is

$E_1^{(0)}, E_2^{(0)}, E_3^{(0)}, \ldots\ldots\ldots E_{n\,(0)}$. eigen function $\psi_1^{(0)}$. $\psi_{2\,(0)}$, $\psi_3^{(0)}$, $\ldots\ldots\ldots \psi_n^{(0)}$.

$$H^{(0)} \psi_{n}{}^{(0)} = E_n^{(0)} \; \psi_n{}^{(0)}$$

Were   n = 1, 2, 3

The eigen function ψ_n for pertirbed system should satisfied equation.

$$H \psi_n = E_n \psi_n$$

Where En = energy eigen value of the modified hamiltanian

Since the perturbation does not as caused large change it means that energy value I waves function for perturbed system will have very small difference from the unperturbed system.

Therefore we can expand energy E and wave function ψ in terms of λ as.

$$E_{\,} = E_{n\,(0)} + \lambda E_{n\,(1)} + \lambda^2 E_n^{(2)} + \ldots$$

$$\psi_n = \psi_{n\,(0)} + \lambda \psi_{n\,(1)} - \lambda^2 \psi_n^{(2)} + \ldots\ldots$$

On substitution (5) and (6) in (4)

$$H^{(0)} + \lambda H^{(1)} \quad {}_\psi{}^{(0)} + \lambda \psi_n^{(1)} + \lambda^2 \psi_n^{(2)} = E^{(0)} + \lambda E^{(1)} + \lambda^2 E^{(2)}$$

$$= {}_\psi{}^{(0)} + \lambda \psi_n^{(1)} + \lambda^2 \psi_n^{(2)}$$

in solving

$$H\psi_n{}^{(0)} + \lambda^n {}^{(0)}\psi_n{}^{(1)} + H^{(1)} \psi_n{}^{(0)} + \lambda^n {}^{(1)}\psi_n{}^{(1)} + H^{(1)} \psi_n{}^{(2)}$$

$$= E_n{}^{(0)} \psi_n{}^{(0)} + \lambda E_n{}^{(0)}\psi_n{}^{(1)} + E_n{}^{(1)} \psi_n{}^{(0)} + \lambda^2 E_n{}^{(0)}\psi_n{}^{(2)} + E_n{}^{(2)}\psi_n{}^{(0)} + E_n{}^{(1)}\psi_n{}^{(0)}$$

If the above   equation is satisfied for all values of λ.

The co-efficient of the same powers of λ on both scales of this equation must be equal.

$$H^{(0)} \psi_n^{(0)} = E_n^{(0)} \psi_n^{(0)}$$

$$\therefore \quad H^{(0)} \psi_n^{(1)} + H^{(1)} \psi_{n\,(0)} = E_n{}^{(0)} \psi_n{}^{(1)} + E_n{}^{(1)} \psi_{n\,(0)}$$

$$\therefore \quad H^{(0)} \psi_n^{(2)} + H^{(1)} \psi_{n\,(1)} = E_n{}^{(0)} \psi_n{}^{(2)} - E_n{}^{(2)} \psi_n{}^{(0)} + E_n{}^{(1)} \psi_{n\,(1)}$$

(9)   represent imperturbed

(10)   1st order porturbation edn.

(11)   2nd order perturbation edn.

In the first order equation the first order energy $E_n{}^{(1)}$ can be evaluated using the expansion theoram using the nirmalisation condition.

$$E_n{}^{(1)} = \int \psi_n{}^{(0)+} H^{(1)} \psi_n{}^{(0)} \, d\tau$$

The first ordedr wave function $\psi_n{}^{(1)}$ can be evaluated using the edn.

$$\psi_n{}^{(1)} = \psi_n{}^{(0)} - \pi \sum_{m=0}^{2} \frac{\int \psi_n{}^{(0)^+} H^{(1)} \psi_n{}^{(0)} d\tau}{E_m{}^{(0)} - E_n{}^{(0)}} X \psi_m{}^{(0)}$$

Where $m \neq n$

and Σ indicates the omission of term

Where m = n

Similarly 2nd order energy $E_n{}^{(2)}$ can be evaluated as,

$$E_n{}^{(2)} = \sum_{m}^{1} \frac{\int \psi_m{}^{(0)^+} H^{(1)} \psi_n{}^{(0)} d\tau \int \psi_m{}^{(0)^+} H^{(1)} \psi_n{}^{(0)} d\tau}{E_n{}^{(0)} - E_m{}^{(0)}}$$

and 2nd order work function

$$\psi n = \sum_{m} \frac{C_m}{E_m{}^{(0)} - E_n{}^{(0)}} \quad c_n{}^{(1)} - \int \psi_m{}^{(0)^+} H^{(1)} \psi_m{}^{(0)} d\tau \, \psi_m{}^{(0)}$$

In degenerate system set of unperturbed wave function has to be determint to ... the perturbed function reduced whom porturbed is removed the correct linear combnation is represented.

$$X_m{}^{(o)} = \sum_{i=1}^{2^{(0)}} kii' \psi m$$

Where $i = 1, 2, 3 \ldots\ldots n$

and $i = 1$

$$X_m{}^{(o)} = kl_1 \psi_{m1}{}^{(0)} + kl_2 \psi_{m2}{}^{(0)} - kl_2 \psi_{m2}{}^{(1)}$$

if $i = 1$

$$X_{m2}{}^{(o)} = k_{11} \psi_{m1}{}^{(0)} + k_{12} \psi_{m2}{}^{(0)} + k_{12} \psi_{m2}{}^{(1)}$$

if $i = 2$

$$X_{m2}{}^{(o)} = k_{21} \psi_{m1}{}^{(0)} + k_{22} \psi_{m2}{}^{(0)} + k_{22} \psi_{m2}{}^{(0)}$$

If X_{m1} is represented as

$$\psi_{nu} = X_{nu}{}^{(0)} + \lambda \psi_{nu}{}^{(1)} + \lambda^2 \psi_{nu}{}^{(2)}$$

and

$$E_{nu} = E_m{}^{(0)} + \lambda E_{nu}{}^{(1)} + \lambda^2 E_{nu}{}^{(2)}$$

perturbed wave equation

$$H\psi_{nu} - E_{nu}\psi_{nu} = 0$$

Substituting the values of $H\psi_{nu}$ and E_{nu} from equation (2), (5) and (6) in equation (7)

$$\left(H^0 + \lambda H^{(1)} + \lambda^2 H^{(2)} + \right)\left(\chi_{nu} + \lambda\psi_{nu} + \lambda^2\psi_{nu} + \right) -$$

$$E_m^{(0)} + \lambda E_{nu}^{(1)} + \lambda^2 E_{nu}^{(2)} +$$

$$\chi_{nu}^{(0)} + \lambda\psi_{nu}^{(1)} + \lambda^2\psi_{nu}^{(2)} = 0$$

On solving we get

$$H\chi_{nu}^{(0)} - E_m^{(0)}\chi_{nu}^{(0)} + \lambda H^{(0)}\psi_{nu}^{(1)} + H\chi_{nu}^{(1)} - E_{nu}^{(0)}\psi_m^{(1)} + E_m^{(1)}\chi_{nu}^{(0)} = 0$$

the 1st order perturbation equation can be obtain by equacting the co-efficient of $\lambda = 0$.

$$\lambda H^{(0)}\psi_{nu}^{(0)} + H^{(1)}\chi_{nu}^{(0)} - E_{nu}^{(0)}\psi_m^{(1)} - E_{nu}^{(1)}\chi_{nu}^{(1)} = 0$$

on expandin $\psi_{nu}^{(1)}$ as,

$$\psi_{nu}^{(1)} = \sum_{m'l'} a_{nu}\,m'l'\,\psi m'l'$$

and substitutina equation (11), (4) in (10) we get,

$$\sum a_{ml\,m'l'}\left[E_{ml}^{(0)} - E_m^{(0)} \right]\psi_{m'l'} = \sum_{i'=1}^{2} k\left[E_{ml} - H(1)\psi_{m'l'} \right]$$

on multiplying equation (12) onboth sides by $\psi^{(0)}_{mj^+}$ andintegrating over all the configuration space we get,

$$\sum_{m'l'} a_{nu\,m'l'}\left[E_{ml}^{(0)} - E_m^{(0)} \right]\int\psi_{mj}^{(0)+}\psi_{m'j'}^{(0)}d\tau =$$

$$\sum_{i'=1}^{2} K\left[E_{ml}^{(1)}\int\psi_{ml}^{(1)}\psi_{nu'}^{(0)+}d\tau - \int\psi_{mj}^{(0)+}H\psi_{m}^{(1)} \right]$$

Since $\psi_{ml}^{(0)}$ and $\psi_{m'l'}^{(0)}$ are orthogonal LHS of the equation is zero. by usng certain symbols.

$$H_{jl'} = \int\psi_{mj}^{(0)*}H\psi_{ml'}d\tau$$

and

$$jl' = \int\psi_{mj}^{(0)*}\psi_{ml}d\tau$$

using these two symbols in equation (13) we will get equation

$$\sum_{l'=1}^{2} n \cdot {}_{il'}{}^{(1)} - {}_{jl} \cdot \varepsilon_{ml} = 0$$

Where j = 1, 2, 3... ∞

Equation (16) represent a system of 2-homogeneous linear simultaneous equations which can be written as,

$$l' = 1...2, \quad j = 1...2$$

$$\left(H_{11}{}^{(1)} - {}_{11}E_{ml}{}^{(1)} \right)k_{11} + \left(H_{12}{}_{(1)} - {}_{12}E_{ml}{}^{(1)} \right)k_{12} + \left(H_{12}{}_{(1)} - {}_{12}E_{ml}{}^{(1)} \right)K_{l2} = 0$$

$$l' = 2$$

$$\left(H_{21}{}^{(1)} - {}_{21}E_{ml}{}^{(1)} \right)k_{11} + \left(H_{22}{}^{(1)} - {}_{22}E_{ml}{}^{(1)} \right)k_{12} + \left(H_{22}{}^{(1)} - {}_{22}E_{ml}{}^{(1)} \right)K_{l2} = 0$$

$$\left(H_{21}{}^{(1)} - {}_{21}E_{ml}{}^{(1)} \right)k_{11} + \left(H_{22}{}^{(1)} - {}_{22}E_{ml}{}^{(1)} \right)k_{12} + \left(H_{22}{}^{(1)} - {}_{22}E_{ml}{}^{(1)} \right)K_{l2} = C$$

We have non-zero solution, the determinant of the co-efficient of the unknown quantities should varnish.

$$\begin{vmatrix} H_{11}{}^{(1)} - {}_{11}E_{ml}{}^{(1)} ... H_{12}{}^{(1)} - {}_{12}E_{ml}{}^{(1)} ... H_{12}{}^{(1)} - {}_{12}E_{ml}{}^{(1)} \\ H_{21}{}^{(1)} - {}_{21}E_{ml}{}^{(1)} ... H_{22}{}^{(1)} - {}_{22}E_{ml}{}^{(1)} ... H_{22}{}^{(1)} - {}_{22}E_{ml}{}^{(1)} \\ H_{21}{}^{(1)} - {}_{21}E_{ml}{}^{(1)} ... H_{22}{}^{(1)} - {}_{22}E_{ml}{}^{(1)} ... H_{22}{}^{(1)} - {}_{22}E_{ml}{}^{(1)} \end{vmatrix} = 0$$

Using the condition

$$jl' = 1, if \quad i = l'$$

and $$jl' = 0, if \quad j \neq l'$$

$$\begin{vmatrix} H_{11}{}^{(1)} - E_{ml}{}^{(1)} & 0 & 0 \\ 0 & H_{22}{}^{(1)} - E_{ml}{}^{(1)} & 0 \\ 0 & 0 & H_{22}{}^{(1)} - E_{ml}{}^{(1)} \end{vmatrix} = 0$$

APPLICATION OF PORTURBATION METHOD :
ground state energy of the He- atom

Application of 1st order perturbation theory of non-degenerate system.

the potential energy for a system of two $\bar{e}$'s and neucleous of charge +Ze

$$V = \frac{-Ze^2}{r_1} - \frac{-Ze^2}{r_2} + \frac{-e^2}{r_{12}}$$

r_1 and r_2 are the distance of $\bar{e} - (1) \, and \, \bar{e} - (21)$ and r_2 is the saperation of $\bar{e}$'s.

The energy repulsion term can be considered as the perturbation of 90 the perturbation hemiltanian

$$H^1 = \frac{e^2}{r_{12}}$$

Hamiltatmian H of wave equation is the sum of unporturbed and perturbed state.

$$H = H^0 + H^1$$

the wave equation for unperturbed Hamiltanian is

$$H^0 \psi^0 - E^0 \psi^0 = 0$$

Which can be solved by taking the sum of two H atom the wave equation two $\bar{e}$ is given by

$$H\psi = \frac{-H^2}{2m_0}(\nabla^2_1 \psi + \nabla^2_2 \quad \psi) + \frac{-Ze^2}{r_1} - \frac{Ze^2}{r_2} + \frac{e^2}{r_{12}}\psi$$
$$= E\psi$$

This equation is applicable to the Li^+ as Be^{+2} etc.

Equation (5) can be written as,

$$H\psi = \frac{-H^2}{2m_0}(\nabla^2_1 + \nabla^2_2) - \frac{-Ze^2}{r_1} - \frac{Ze^2}{r_2}\psi + \frac{e^2}{r_{12}}\psi = E\psi$$

$$H\psi = \frac{-H^2}{2m_0}(\nabla^2_1 + \nabla^2_2) - \frac{-Ze^2}{r_1} - \frac{Ze^2}{r_2}, \qquad H^1 = \frac{e^2}{r_{12}}$$

The unporturbed equation can be substituted

$$\psi_0 = \psi^0_1 \psi_{0_2} \text{ and corresponding energy can be written}$$
$$E^0 = E^0_1 E_{0_2}$$

The equation for ψ^0_1 is written as

$$\nabla^2_1 \quad \psi^0_1 + \frac{2m_0}{H^2}E^0_1 \quad + \frac{Ze^2}{r_1}\psi^0_1 \quad = 0$$

for ψ^0_2.

$$\nabla^2_2 \quad \psi^0_2 + \frac{2m_0}{H^2}E^0_2 + \frac{Ze^2}{r_2}\psi^0_2 = 0$$

The equation (6) and (7) are Hydrogen like wave equation and the solution for the energy is given as,

$$E\,_2^{\,0} = E\,_1^{\,0} = \frac{-Z^{\,2}WH}{n^{\,2}}$$

Where $\qquad W_H = \dfrac{2\pi^2 m_0 e^4}{H^{\,2}} = \dfrac{m_0 e^4}{2H^2}$

Thus the total unperturbed energy is given by

$$E^0 = -\,2Z^{\,2}WH \quad \text{for n = 1.,}$$

Ground State the total perturbed energy E^1 is the average value of the perturbed function

$$H^{\,1} = \frac{c^2}{r_{12}} \quad \text{over the unperturbed state, of the system which is given by,}$$

$$E^1 = \int \psi^{0'} \, H^{\,1} \, \psi^0 \; d\tau = \int \frac{c^2}{r_{12}} \left|\psi^0\right|2 \; d\tau$$

$$\psi^0 = \psi^0_1 \psi^0_2 = \sqrt{\frac{Z^3}{\psi a_0^{\,3}}} \; e^{-\delta 1/2} \sqrt{\frac{Z^3}{\psi a_0^{\,3}}} \; e^{-\delta 2/2}$$

Where $\delta = \dfrac{2Zr}{a_0}$, $\quad a_0 = \dfrac{-H^2}{m_0 e^2}$

hence, $\qquad \psi_0 = \dfrac{Z^3}{\pi a_0^{\,3}} \; e^{\frac{-\delta 1 - \delta 2}{2}}$

The volume element $d\tau$ in (r,θ,φ) co-ordinate
is $d\tau = r_{12}\ dr_1 \sin\theta_1\ d\theta_1\ d\theta_1\ r_2\ \sin\theta_2\ d\theta_2\ d\theta_2$,

$$\therefore E^{\,1} = \frac{Ze^2}{32\pi_2 a_0} \int_0^{2\pi} \int_0^{n2} \int_0^{} \int_0^{2\pi} \int_0^{n2} \int_0^{} e^{\frac{-\delta 1 - \delta 2}{\delta 12}} \delta_1^{\,2}\, d\delta_1 \sin\theta_1 \cdot d\theta_1 \cdot d\theta \cdot \delta_2 \sin\theta_2\, d\theta_2\, d\theta_2$$

Where $\delta_{12} = \dfrac{2Zr_{12}}{a_0}$

The above integral represent the mutual electrostatic energy of 2 spherically symmetrical distribution of electricity with density function. $e^{-\delta 1}$ and $e^{-\delta 2}$ respectively.
by solving equation (14) we get,

$$E^1 = \frac{5}{4}\, ZWH$$

Since $\qquad E = E^0 + E^1$

$$= -2Z^2 WH + \frac{5}{4} ZWH$$

$$= -2Z^2 - \frac{5}{4} ZWH$$

for Hellium $\quad Z = 2$

$$E = -(2 \cdot 4 - 5/4 \cdot 2)\, WH$$

$$= \frac{-11}{2}\, WH$$

Since $\quad WH = \dfrac{m_0 e^4}{2H^2}$

$$E = \frac{-11}{2} \cdot \frac{m_0 e^4}{2H^2} = \frac{-11}{2} \cdot \frac{m_0 e^4}{H^2}$$

is ground state energy of He atom by applying the 1st order pertubation non-degenerate system.

Variation Method :

Method of linear combination C perticular form of variation method.

$\downarrow$

LCAO approximation

$\wedge$

VB mo

Whoch find application for the combination of atomic wave function.

ψ is having the characteristics of ψ_1 and ψ_2 complete wave function

$$\psi = C_1 \psi_1 + C_2 \psi_2 \text{(can be writen as linear combination)}$$

C_1 and C_2 are co-efficient chosen in such a way that energy is minimized.

Principle of linear combination states that if a number of different approximate wave function are solutions of to the true wave function, then any linear combination of these wave function is also a solution. thus for a set of function ψ_1, $\psi_2 \ldots \psi_n$

$$\psi = C_1 \psi_1 + C_2 \psi_2 + C_3 \psi_3 + \ldots\ldots\ldots +$$
$C_n \psi_n$ for atomic wave function to combine

(1)   Should have the same energy ψ_1, $\psi_2 \ldots \psi_n$

(2)   should have the same symmetry

(3)   Should overlap or combine effectively (appreciably)

nuclei   A   5

$$\psi_A + \psi_B \qquad \text{at wave function}$$

$\psi = C_A \psi_A + C_B \psi_B$ according to principle of linear combination can be written as,

$$\psi = N(\psi_A + \psi_B) \qquad \text{indicates the contribution of } \psi_B \text{ towards } \psi.$$

N = Normalized factor

by neglacting N

$$\psi = \psi_A + \lambda \psi_B$$

approximation called LCAO

- LCAO approximation to 'n' atomic orbitals,

if φ_1 , φ_2 , φ_3 etc are wave function

$$\psi = C_1 \varphi_1 + C_2 \varphi_2 + C_3 \varphi_3 + \dots \dots C_n \varphi_n$$

to begin with we can consider the wave function ψ as a combinatin of only two variables.

$$\psi = C_1 \varphi_{21} + C_2 \varphi_2$$

can be consider as trial wave function.

to corresponding energy E can be calculated.

$$E = \int \psi^* H\psi \, d\tau \,/\, \psi^* \psi \, d\tau$$

Substituing for ψ

$$E = \frac{\int [C_1\varphi_1 + C_2\varphi_2]^* \, H[C_1\varphi_1 + C_2\varphi_2] \, d\tau}{\int [C_1\varphi_1 + C_2\varphi_2]^* [C_1\varphi_1 + C_2\varphi_2] \, d\tau}$$

by solving

$$E = \frac{\int [C_1\varphi_1 + C_2\varphi_2]^* [C_1\varphi_1 + C_2\varphi_2] \, d\tau}{}$$

$$= \int [C_1\varphi_1 + C_2\varphi_2]^* \, H[C_1\varphi_1 + C_2\varphi_2] \, d\tau$$

By saving we get,

$$\therefore E = C_1^2 \int \varphi_1 \varphi_1 \, d\tau + 2C_1C_2 \int \varphi_1 \varphi_2 \, d\tau + C_2^2 \int \varphi_2 \varphi_2 \, d\tau$$

$$= C_1^2 \int \varphi_1^* H \varphi_1 \, d\tau + 2 C_1 C_2 \int \varphi_1^* H \varphi_2 \, d\tau + C_2^2 \int \varphi_2^* H \varphi_2 \, d\tau$$

to minimized E.

$$\frac{\partial E}{\partial C_1}\bigg|_{C_2} = \frac{\partial E}{\partial C_1}\bigg|_{C_2} = 0$$

equation (5) written to C_1 keeping C_2 constant.. we get,

$$E = 2C_1 \int \varphi_1 \varphi_1 \, d\tau + 2C_2 \int \varphi_1 \varphi_2 \, d\tau + \frac{\partial E}{\partial C_1} C_1 \int \varphi_1 \varphi_2 \, d\tau$$

$$+ \frac{\partial E}{\partial C_1} 2C_1 \int \varphi_1 \varphi_2 \, d\tau - C_2^2 \int \varphi_2 \varphi_2 \, d\tau =$$

$$2C_1 \int \varphi_1^* H \varphi_1 \, d\tau + C_2^2 \int \varphi_2^* H \varphi_2 \, d\tau + 2C_2 \int \varphi_2^* H \varphi_2 \, d\tau$$

an equivalent equation will be obtain

for $\quad C_2 \quad \dfrac{\partial E}{\partial C_2} \quad C_1$

now to simply to equation certain terms of symbol can be introduced.

$$H_{ij} = \int \varphi_i^* H \varphi_j \, d\tau. \qquad S_{ij} \int \varphi_i^* \varphi_j \, d\tau$$

i and j of orbitals

by applying the condtions

$$\frac{\partial E}{\partial C_1} = 0 = \frac{\partial E}{\partial C_2} \quad \text{to equation}$$

We can write (6) as

$$(H_{11} - E S_{11}) C_1 + (H_{12} - E S_{12}) C_2 = 0$$

and $\dfrac{\partial E}{\partial C_2} = 0$ an eqiovalent equation can be obtain.

$$(H_{21} - E S_{21}) C_1 + (H_{22} - E S_{22}) C_2 = 0$$

secular equation

$$\begin{vmatrix} H_{11} - E S_{11} & H_{12} - E S_{12} \\ H_{21} - E S_{21} & H_{22} - E S_{22} \end{vmatrix} = 0$$

It is called secular determinatin.
this is in case of two variables

$$\psi = C_1\, \varphi_1 + C_2\, \varphi_2$$

If we express ψ in terms of 'n' independent variables

$$\begin{vmatrix} H_{11} - ES_{11} & H_{12} - ES_{12} & H_{1n} - ES_{1n} \\ H_{21} - ES_{21} & H_{22} - ES_{22} & H_{2n} - ES_{2n} \\ H_{n1} - ES_{n1} & H_{n2} - ES_{n2} & H_{nn} - ES_{nn} \end{vmatrix} = 0$$

Since this equation (11) is nth order in principale it gives n roots. E_1, E_2, E_3 .... E_n and therefore set of 'n' wave function

$$\psi_1, \psi_2, \psi_3 \ .... \psi_n$$

'n' atomic orbital combine gives 'n' molecular orbital, it gives 'n' energy state.

Summerise :

(1)   Application of variction principle and LCAO approximation to 'n' atomic orbital system will give 'n' secular equation.

(2)   Since the co-efficients of the secular collection have non-zero values acceptable solutions are possible. only when the secular cleterminant itself is zero.

(3)   The dimensions of secular determinant will be n x n

- The form of secular determinant will give perticular value of E system corresponding to the molecular wave function of the orbitals these values can be incertain inserted in secular equation to get the value of co-efficient.

Significance of integrall H and S :

(1) H_{11} or α coulonbic intergral and it gives the energy of an electron in an atomic orbital. (energy 2 is -ve)

(2) H_{12} or β represent the resonance or exchange energy within 1 and 2 e and hence called resonance integrall or exchange integral β represent the energy of

$\bar{e}$ when it is between two atms this is also -ve quantity.

(3) S_{12} Known as overlap integral of measures the probability of finding $\bar{e}$ on 1 and 2 when two orbitals are seperate from each other $S_{11} = 0$ but during overlap S_{12} will have d high value.

In order to have a moleclar orbital through LCAO approximation the atomic orbital must have the same enrgy. Same symmetry should overlap (combine) appreciably.

Thus the LCAO approximation leads to creation of several molecular orbitals to each

molecule which is basis of mo. theory.

LCAO approximation :

VB   MO

VB : overlap of atomic orbits

$\bar{e}$ in the atomic orbitals remains in the same obbitals even after the bond formation.

- The $\bar{e}$ pairs are localized between two atom.
- From this theory we can know to shape of molecule.

MO : Atomic orbital combine to linearly to from a molecular orbital and $\underset{-}{e}$ are filled in the

MOs starting from lowest energy M.O. with $2\ \bar{e}$ each

∴ e are delocalised

∴ it is not used for shape of molecule.

∴ the magnatic, optical, electrical property we can use the M.O. theory.

EXPLAIN VALENCE STALE AND VALENCE STALE SONISATION POTENTIAL (VSIP)

The valence stale of an atom for a given molecular electronic stale is the stale in which the atom exists in the molecule. Since individual atoms do not really exist in the molecules the valence state concept is an approximate one.

We use the VB wave function define the valence stale of an atom as the wave function obtained on removing all other atom infinity, while keeping the form of molecular wave function inhalant . This process is purely hypothetical and the valence stale is not in general a stationary atomic state.

Consider an atom in which electrons are placed in suitably hybridized orbitals. The atom is then said be promoted a valence stale . In otherwards, the valence stale of an atom is a stale which is prepared through electronic rearrangement ____ the bond formation e.g. In the localized bond model a carbon atom in a conjugalid system is supposed ____ be in represented by the configuration.

$$1s^2 \, (tr_1)^1, \, (tr_2)^1, \, (tr_3)^1 \, (2r_2)^1 \, \ldots\ldots\ldots \, (1)$$

where tr_1, tr_2, tr_3 are trigonal hybrinds formed from 2s, 2px & 2py orbitals. One may assume without any serious error that the configuration of a carbon atom given by (1) is equivalent that given by

$$1s^2, \, (2_s)^1 \, (2p_x)^1 \, (2p_y)^1 \, (2p_z)^1 \, \ldots\ldots\ldots(2)$$

because three electrons occupying the hybud orbitals that have 1/3s and 2/3p characters should be equivalent ____ one electron occuping the s- atomic orbital and two p-atomic orbitals.

Thus the valance stale Ionization Potential (VSIP) of $2p_2$ orbital in carbon is given by the difference in energy between neutral carbon in the configuration $1s^2$, $(2_s)^1$ $(2p_x)^1$ $(2p_y)^1$---(3) similary the VISP of 2s-orbital in carbon is given by the energy difference between configuration (1) of neutral C atom and configuration of C^+ $1s^2$, $(2_s)^1$ $(2p_x)^1$ $(2p_y)^1$ $\ldots\ldots\ldots(4)$

The VSIP for 2_p electron in an-Sp^3 hybridized c. atom is the energy difference between the valence stale of Sp^3 hybudiled C and the valance stale q Sp^2 hybudized C^+ jb 2s hybridization is included in the 2p bonding oxygen orbitals of water the oxygen valence stale is formed G be a linear combination involving ferms of the configuration $1s^2 \, 2s^2 \, 2p^4$, $1s^2 2s^1 2p^5$. $1s^2 \, 2p^6$ Hybrudugatuib gives a mixing of configurations S in the valence stale.

The most commonly used values of VISP in Extended Huckel Theory) are

- 13.0 ev for hydrogen 1s
- 21.4 ev for carbon 2s
- 15.4 ev for carbon 2p
- 26.0 ev for nitrogen 2s
- 13.4 ev for nitrogen 2p
- 32. 0 - 35.3 ev for oxygen 2s
- 14.8 – 17.8 ev for oxygen 2p

The VSIP are used estimate integrals in the semi empirical calculations eg. ppp method and in EHT

PPP   PERISER PARR POPLE

EHT   EXTENDED HUCKEL THEORY

Lader Operaters (Step –up and Step-down Operators)

i) $[L^2, L_{\pm}] = 0$

ii) $[L+, L_z] = -hL+$ (Pooerties of Ladder operators)

iii) $[L + L-] = -2hL_z$

Lader operators are known as raising and lowering operators. These operators are mentioned below.

$$L\hat{}+ = L_x + iL_y$$

$$L\hat{}_- = L_x - iL_y$$

These operators are increased & decreased the Eigen value. So they are known as raising and lowering operators. Properties of these operators are that they commute with L^2 operators but they do not commute with each other and L_z. These properties are mentioned below.

(i) $[L^2, L_+]$ $= [L^2, (L_x+iL_y)]$

                              $= L^2 (L_x + iL_y) - (L_x + iL_y) L^2$

                              $= L^2 L_x + iL^2 L_y - L_x L^2 - iL_y L^2$

                              $= L^2 L_x - L_x L^2 + iL_2 L_y - iL_y L^2$

                              $= [L_2, L_x] + i[L_2, L_y]$

                              $= 0 + i(0)$ (There for L^2 is commute with L_x, L_y, L_y)

        $\therefore [L^2, L_+] = 0$

By this way $[L^2, L_{\pm}] = 0$ (that means $L_{\pm}$ commute with L^2)

There for $L_2, L_{\pm} = 0$

(ii) $[L+, L_z]$ $= [(L_x + iL_y), L_z)]$

                              $= [L_x + iL_y) L_z - L_z (L_x + iL_y)$

$$= L_x L_z + iL_y L_z - L_z L_x - iL_z L_y$$

$$= [L_x, L_z] + i [L_y, L_z]$$

$$= ih/2\pi L_y + i \times (+ i\text{-}h/2\pi L_x)$$

$$= h/2\pi (L_x + iL_y)$$

$$[L_+, L_z] \qquad = h/2\pi L_+$$

By this way $[L_+, L_z] = -h/2\pi L\text{-}$

$$[L_\pm, L_z] \qquad = h/2\pi L_\mp \qquad (OR) \qquad [L_z, L_\pm] = h/2\pi L_\pm$$

(iii) $\quad [L_+, L_-] = [(L_x + iL_y), (L_x - iL_y)$

$$= (L_x + iL_y)(L_x - iL_y) - (L_x - iL_y)(L_x + iL_y)$$

$$= (L_x{}^2 - iL_xL_y + iL_yL_x - i^2Ly^2) - (L_x{}^2 + iL_xL_y - iL_yL_x - i^2L_y{}^2)$$

$$= \cancel{L_x{}^2} - iL_xL_y + iL_yL_x + \cancel{L_y{}^2} - \cancel{L_x{}^2} - iL_xL_y + iL_yL_x - \cancel{L_y{}^2}$$

$$= -2i\, iL_xL_y + 2iL_yL_x$$

$$= -2i (L_xL_y - L_yL_x)$$

$$[L_+, L_-] \qquad = -2i \times ih/2\pi L_z \qquad = 2h/2\pi L_z$$

Eigen value & Eigen function of $L^{\wedge 2}$ & $L^{\wedge}_z$

if we mentione ϕ_j, m, two equation of Eigen value describe below

$$L^{\wedge 2} \phi_j, m = Kj\, \phi_j, m \quad ....(1)$$

$$L_z \phi_j, m = Kmj\, \phi_j, m \quad ....(2)$$

where, j & m are quantum numbers

$$L^{\wedge}_z \cdot L^{\wedge}_z \cdot \phi_j, m = L^{\wedge}_z (Km\phi_j, m)$$

$$L^{\wedge 2}_z \phi_j, m = Km^2 \cdot \phi_j, m \quad ....(3)$$

$$(L^2 - L^{\wedge 2}_z) \phi_j, m = (Kj - Km^2)\, \phi_j, m$$

OR $(\hat{L}_x^2 + \hat{L}_y^2)\, \phi_{j,m} = (Kj - Km^2)\, \phi_{j,m}$ ....(4)

$(\hat{L}_x^2 + \hat{L}_y^2)$

$\hat{L}_x^2 + \hat{L}_y^2$

i.e. Km value $= Kj - Km^2 \geq 0$

$Kj \geq Km^2$

Why they are step up down

Now we see what the effect of lader operators on eigen value and why they known as step up and step down.

$\hat{L}^2 L + \phi_{j,m} = L^\wedge + L^2 \phi_{j,m} = L^\wedge + (Kj\phi_{j,m})$

$L^2 (L + \phi_{j,m})$ $\qquad = Kj (\hat{L}_+ \phi_{j,m})$

Now, what the effect of L+ on km, for prove it add $\hat{L}^\wedge + L_z$

$\hat{L}_+ L_z \phi_{j,m} \quad = \hat{L}_z \hat{L}_+ \phi_{j,m} = L[\hat{L}^\wedge + \hat{L}_z + \hat{L}_z L_+^\wedge - L_+ \hat{L}_z] \phi_{j,m}$

$\qquad = \{ \hat{L}^\wedge + \hat{L}_z + [\hat{L}_z, L_+]\} \phi_{j,m}$

$\qquad = \{ L_+ L_z \phi_{j,m} + [\hat{L}_z, L_+]\phi_{j,m}\}$

$\qquad = L_+ km\phi_{j,m} + h/2\pi L_+ \phi_{j,m}$

$\qquad = L_+ (km + h/2\pi)\, \phi_{j,m}$

$\qquad = (km + h/2\pi)\, \hat{L}_+ \phi_{j,m}$

$\hat{L}_z L_+ \phi_{j,m} \quad = (km + h/2\pi) \hat{L}_+ \phi_{j,m}$

on this equation we prove it L_z is eigen function of $L_+ \phi_{j,m}$

L- operator does not effect on $\hat{L}^2$ but effect on eigen value of L_z

$\hat{L}^2 (L - \phi_{j,m}) \quad = kj\, (\phi_{j,m})$

$\hat{L}_z (L - \phi_{j,m}) \quad = (km - h/2\pi)\, (L - \phi_{j,m})$

$\hat{L}^2 (L \pm \phi j,m) = kj (\phi j,m)$

$\hat{L}z (L \pm \phi j,m) = \frac{km+h/2\pi}{km-h/2\pi} \phi j,m$

Now, L+ & L- are not hermition but L+L- & L-L+ are hermition

$\hat{L}_+ = \hat{L}_x + i\hat{L}_y$

$\hat{L}_- = \hat{L}_x - i\hat{L}_y$

$< \Psi_i / L_+ / \Psi_j > = < \Psi_j / L_+ /\Psi_i>^* ....(A)$

$< \Psi_i / L_+ / \Psi_j> = < \Psi_i / L_x / \Psi_j> + i < \Psi_i / L_y / \Psi_j>$

$< \Psi_i / L_+ / \Psi_j> = < \Psi_j / L_x / \Psi_i>^* + < \Psi_j / L_y / \Psi_i> ....(1)$

$< \Psi_j / L_+ / \Psi_i>^* = < \Psi_j / (L_x + iL_y) / \Psi_i >^* + < \Psi_j / L_x / \Psi_i >^* - i< \Psi_j / L_y / \Psi_i >^* .....(2)$

from equation (1) & (2)

$< \Psi_i / L_+ / \Psi_j> \neq < \Psi_j / L_x / \Psi_i>^*$

Therefore, L+ is not hermition

$\left. \begin{array}{l} L_+L_- = \hat{L}_x^2 + \hat{L}_y^2 + \hat{L}_z \\[2mm] L_-L_+ = \hat{L}_x^2 + \hat{L}_y^2 - \hat{L}_z \end{array} \right\} \quad (A)$

$\hat{L}_x , \hat{L}_y , \hat{L}_z$

$< \Psi_i / L_+ L_- / \Psi_j> = < \Psi_j / L_+ L_- / \Psi_i>^* ....(B)$

$< \Psi_i / L_+ L_- / \Psi_j> = < \Psi_i / \hat{L}_x^2 / \Psi_j> = < \Psi_i / \hat{L}_y^2 / \Psi_j> = < \Psi_i / \hat{L}_z^2 / \Psi_j> .....(1)$

$< \Psi_i / L_+ L_- / \Psi_j> = < \Psi_j / \hat{L}_x^2 / \Psi_i> = < \Psi_j / \hat{L}_y^2 / \Psi_i> = < \Psi_j / \hat{L}_z / \Psi_i> .....(2)$

$< \Psi_j / L_+ L_- / \Psi_i>^* = < \Psi_j / \hat{L}_x^2 / \Psi_i>^* + < \Psi_j / \hat{L}_y^2 / \Psi_i>^* = < \Psi_j / \hat{L}_z / \Psi_i> .....(3)$

$\qquad\qquad\qquad\qquad\qquad = < \Psi_j / \hat{L}_x^2 + \hat{L}_y^2 + \hat{L}_z / \Psi_i >^*$

$$= <\Psi_j \,/\, \hat{L}{+}\hat{L}{-}\,/\, \Psi_i> * \ldots(4)$$

Rights side of equation (2) & (4) are equal so the left side will be equal

$$< \Psi_i \,/\, L{+}\, L{-}\,/\, \Psi_j> \;=\; < \Psi_j \,/\, L{+}\, L{-}\,/\, \Psi_i> *t$$

therefore L+L- both are hermition and this way we prove it L.L+ are hermition.

–

www.ingramcontent.com/pod-product-compliance
Lightning Source LLC
Chambersburg PA
CBHW080822180526

45168CB00006B/2544